Project 11356/M
Guided Missile Frigate

Black Sea Guardian/Indian Sentinel

Hugh Harkins

Project 11356/M
Guided Missile Frigate

Black Sea Guardian/Indian Sentinel

© 2021 Hugh Harkins FRAS, MIstP, MRAeS

Centurion Publishing

United Kingdom

ISBN 10: 1-903630-13-4
ISBN 13: 978-1-903630-13-6

This volume first published in 2021

CONTENTS

INTRODUCTION

The Project 11356/M Guided Missile Frigate/Guard Ship was initially developed by Severnoye Design Bureau in Russia to meet an Indian Navy requirement that emerged as the Talwar and Teg Class respectively. The basic design, which was, in 2009, selected for service with the Russian Federation Black Sea Fleet as the Project 11356M, could conduct the full spectrum of Frigate related missions – anti-air warfare, anti-submarine warfare and surface strike etc., courtesy of advanced mission sensors and weapon systems.

The volume sets out to detail the Project 11356/M design, sensors, incorporated weapon systems and entry into service with the Indian and Russian Federation navies. All technical data relating to the respective ship platforms, sensors, weapon systems and components have been furnished by the respective design bureaus, builders and operators, as has the imagery/graphics.

1

PROJECT 11356/TALWAR & TEG CLASS FRIGATE DESIGN ORIGINS

The Project 11356 design was developed as a general purpose Guided Missile Frigate to replace outmoded Soviet era warships, primarily for export potential to meet an Indian Navy requirement for such vessels that emerged as the Talwar Class. The new Frigate design, which enhanced the Russian Federation's position in the competitive export market, was further developed into the Teg Class for the Indian Navy as that service embarked upon a program to vastly increase its combat capabilities in a quasi-arms race with what she viewed as her traditional regional competitors, Pakistan and China (India and China shared a contested land border, but did not share a maritime border). India was also faced with the increasing presence of NATO (North Atlantic Treaty Organisation) and NATO partner nation warships on her periphery as the Arabian Sea and Indian Ocean became increasingly militarised.

An enhanced capability development of the Project 11356 was selected by the MODRF (Ministry of Defence of the Russian Federation) to meet a Guard Ship (General Purpose Frigate) requirement for the Russian Navy Black Sea Fleet. This was done in the background of a large scale modernisation of the Russian Federation armed forces from the late 2000's, to meet what she perceived as an increasing threat from a NATO encroaching on her western and southern land and maritime borders. NATO had become increasing hostile toward Russia in the wake of Russia's defence of an internationally recognised/authorised peacekeeping operation in the Georgian breakaway Republics of South-Ossetia and Abkhazia that was attacked with a massive bombardment of the civilian population centre of Tskhinvali by Georgian armed forces in August 2008. While the Russian defence of the peacekeeping operation, and the civilian population, was falsely portrayed as an invasion of Georgia by western media and political establishments, it was, in actuality, limited to neutralising Georgian military units attacking Russian Peacekeepers, allied forces and civilian population centres. An internal European Union led International investigation laid the blame for the war at the Georgian governments' feet (Independent International Fact Finding Mission on the Conflict in Georgia, 2009).

Despite this, the scene had been set for an increased NATO alliance naval and air presence in the Black Sea as NATO viewed the Russian operations in the region a threat to the alliances attempt at regional dominance (this included the longer term goal to incorporate Georgia into the alliance). This was viewed by Moscow as a direct threat to her southern borders, resulting in a significant modernisation and increase in capability of her regional surface and subsurface fleets, air power fleets and coastal defence forces to counter the increased NATO activity, which, ignoring the facts on the ground, has, in the second and third decades of the twenty first century, fueled the false Russian invasion narrative, which was regularly resurrected by western mainstream media and politicians.

The most prominent of the surface vessel designs ordered for the Black Sea Fleet was the three Project 11356M (11356.20) Guards Ships that were commissioned in 2016-2017. These vessels, the *Admiral Grigorovich*, *Admiral Essen* and *Admiral Makarov*, transformed the anti-air warfare, anti-submarine warfare and, along with the Project 636.3 submarines and other small size surface warships, the anti-ship and long-range land attack capabilities of the Black Sea Fleet, which also took on the main burden of providing ships for the Russian Permanent naval detachment in the Mediterranean. This included launching Kalibr-NK land attack cruise missiles at ISIL (Islamic State of Iraq and the Levant) targets in Syria in support of the Russian Federation military operation to support the Syrian government forces in their fight against the extremist organisation in the second half of the second decade of the twenty first century.

The Project 11356M Guard Ships (*Admiral Grigorovich* illustrated) would be the first new design of large surface warship class ordered for the Russian Navy since the dissolution of the Soviet Union in December 1991. Yantar Shipyard

The Project 11356 (Indian Talwar & Teg Class) was developed from the Soviet Project 1135 Burevestnik Class Anti-Submarine Warfare/Guard Ship, designed and developed in the 1970's. This design (NATO Reporting name Krivak III) was itself developed as an alternative to the large anti-submarine warfare ships of the Project 1134A/1134B types, with essentially the same armament systems as their larger cousins, but with reduced ammunition capacity. The Project 1135, designed by N.P. Sobolev, had incorporated a number of novel traits identifiable in modern designs, including a gas turbine main power plant, incorporating a cruise reduction gearbox capable of driving two propellers from a single engine (Severnoye).

The basic design of the Project 11356 was a deep development of the Project 1135 Burevestnik Class Anti-Submarine Warfare/Guard Ship, designed and developed in the 1970's. This design (NATO Reporting name Krivak III) was initially designed and built for the Soviet Navy – The Russian Navy ship *Pytlivy* is illustrated. MODRF

The Project 11356 was significantly larger that the Project 1135. The latter had a full load displacement of 3305 tonne; length, 123 m; beam, 14.2 m and draught, 4.57 m. The ships could generate a speed of 30-32 knots (~55.56-59.3 km/h), with a cruising range of ~6437 km at a speed of 14 knots (~25.92 km/h) and ~1014 km at 30 knots. Project 1135 ships were armed with an Osa-M short-range surface to air missile system with 20 missiles, 100 mm calibre AK-100 gun mounting, an RBK Blizzard (URPK Trumpet) anti-submarine system, two twin torpedo tube mountings for 4 x 533 mm anti-submarine torpedoes, two RBU-6000 12-tube anti-submarine rocket launchers and 20 mines (MODRF).

The basic design of the Project 11356 was completed in the late 1990's under designer V.A. Perevalov. The Indian Navy, for which, as noted above, the new Frigate class was initially developed, ordered two separate batches of three Project 11356, which were delivered in 2003-2004 (first batch) and 2012-2013 (second batch) (SDB). The MODRF authorised construction of six Project 11356M for the Russian Navy in August 2009 (Severnoye, 2013a).

Previous page top: Artist rendering showing a pair of Project 11356 derived Talwar or Teg Class Frigates in an anti-air warfare environment, the background vessel launching a 9M317E surface to air missile from the Shtil-1 single-rail rapid fire launch system. Previous page bottom: A Talwar Class Frigate, ordered for the Indian Navy, during sea trials under the Russian Federation flag. This page: A Project 11356 vessel is prepared for launch at the Yantar Shipyard, Kaliningrad, Russia, with another such ship in the background (top) and Project 11356 vessels fitting out at the Yantar Shipyard (bottom). Rosoboronexport/Yantar Shipyard

In both Indian and domestic Russian Federation service, the Project 11356/M was intended to be capable of combating adversary sub-surface and surface warships in various maritime environments, ranging from deep ocean to littoral. The design was also intended to conduct long-range strike against land targets and counter adversary airborne threats, whilst operating in multi-warship task groups or independently, and to conduct maritime trade escort/protection and, in regard to the Russian Federation, guard ship operations to protect the integrity of Russia's maritime borders in war or non-war conditions (Severnoye, 2018).

Project 11356M Guard Ships *Admiral Essen* **(top) and** *Admiral Makarov* **(bottom).** MODRF/Rosoboronexport

PROJECT 11356/M GUIDED MISSILE FRIGATE/GUARD SHIP

The basic characteristics of the Project 11356 included a length of 124.8 m; beam, 15.2 m; draft, 4.2 m; a standard displacement of 3620 tonne and a full load displacement of 4025 tonne. The ships main power plant consisted of a UGT gas turbine rated at 41200 kW (~56,043 hp.) output, driving two propellers. The engines bestowed upon the design a maximum speed of 30 knots (~55.56 km/h) ±0.5 knots (~0.92 km/h) and an economical cruise speed of 14 knots (~25.92 km/h). Range at economical cruising speed was ~7805 km (4850 miles) and endurance for the ships, with a standard compliment of 193 (Russian domestic service), was set at 30 days. The primary electrical power supply was provided by 4 x WCM800/5 diesel alternators, with a combined power output of 3200 kW (Severnoye).

To fulfill the mission requirements the Project 11356 was equipped with a plethora of advanced sensors and weapon systems. The radio-electronic suite included a single Fregat-M2EM radar complex designed for detection and tracking of airborne and surface targets, a 3Ts-25E target acquisition radar, MR-212/201-1 short-range navigation radar and a Trebovanie-M action information system. The sonar suite consisted of one MGK-335EM-03 sonar complex (an alternative was the MGK-335EM-02 Sonar complex) and one Vinyetka-EM sonar complex with flexible towed antenna array (Rosoboronexport & Severnoye). The navigation suite consisted of a Ladoga-ME-11356 inertial navigation and stabilisation system, an induction log, a navigation echo sounder, 2 x magnetic compass and a dead-reckoning and plotting system (Rosoboronexport).

The vessels incorporated a comprehensive weapon suite to enable them to conduct operations against sub-surface, surface and airborne targets. For operations against surface targets the primary armament consisted of a complex for the launch of Kalibr-NK cruise missiles in domestic Russian Federation ships or the Club-N complex on ships intended for export. The primary gun armament consisted of a single A-190/E-5P-10/E 100 mm calibre gun/fire control complex and two AK-630M 30 mm automatic gun mounts. For engagements against sub-surface targets the Project 11356/M was armed with a single RBU-6000 anti-submarine rocket projector and two DTA-53-11356 twin torpedo tubes for launching 4 x SET-65E/53-65KE anti-submarine warfare torpedoes. In the air warfare role the Project 11356 was armed with a Shtil-1 surface to air missile complex and could be armed with a Kashtan-M surface to air missile/automatic gun complex. The defensive systems included a PK-10 short-range jamming complex, incorporating four KT-126 launchers for 120 decoys. The aviation complement consisted of a single helicopter, either a Russian Helicopters Ka-27, Ka-28 (export) or Ka-31, accommodated in a hanger complex ahead of the aft section flight deck (Rosoboronexport).

The first Project 11356M Guard Ship built for the Russian domestic fleet was the *Admiral Grigorovich*, **officially referred to as a pre-series vessel.** Yantar Shipyard

Project 11356 General Characteristics – data furnished by Severnoye Design Bureau, United Shipbuilding Corporation (USC) & Rosoboronexport

Propulsion: UGT gas turbine rated at 41200 kW output (~56,043 hp.) (USC)
Propellers: 2
Electrical power supply: 4 x WCM800/5 diesel generators with a combined power output of 3200 kW
Basic dimensions: Length, 124.8 m; beam, 15.2 m and draft, 4.2 m (standard displacement) or 4.66 m (full displacement)
Displacement: 3620 tonne standard and 4035 tonne full load (Rosoboronexport) (Severnoye Design Bureau states 3860 tons for Talwar Class) = to ~3502 tonne
Maximum speed: 30 knots ± 0.5 knots at ambient temperature of 15° C (Centigrade), reducing to 28 knots at ambient temperature of 40° C
Economical cruise speed: 14 knots
Range: ~7805 km (4,850 miles) at economical cruising speed (Severnoye Design Bureau states ~8041 km (5000 miles))
Endurance: 30 days
Complement: 193 (Severnoye Design Bureau states 220 for Talwar Class)

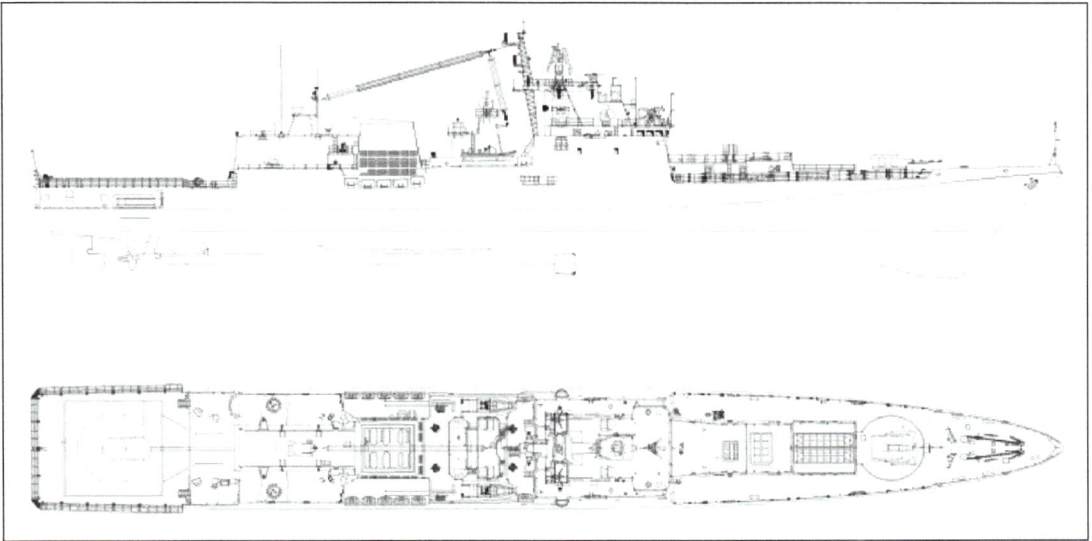

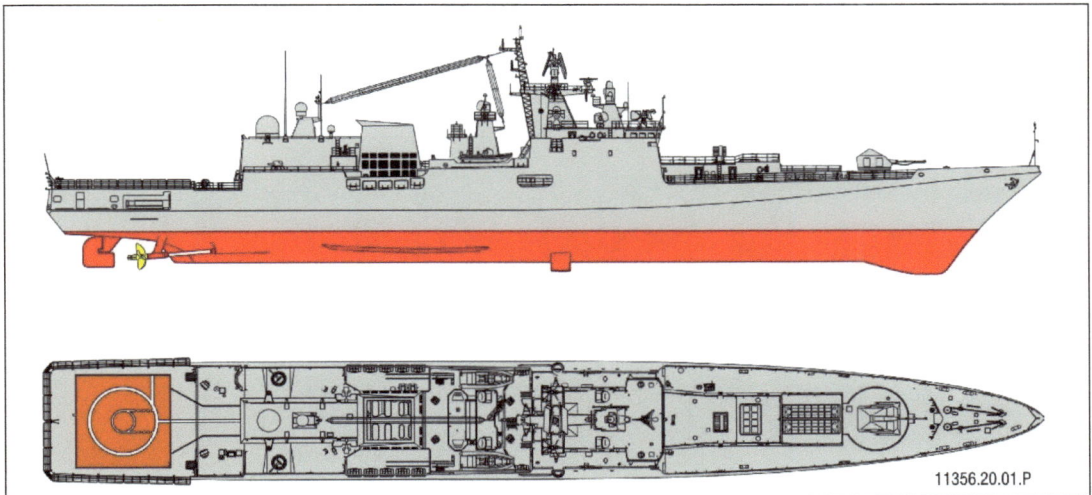

Transparent (top) and detailed (centre) starboard side-on and upper plan view of the general layout of the Project 11356M (11356.20.01.P), which was also representative of the Project 11356 Talwar Class and Teg Class built for the Indian Navy. Bottom: The second Project 11356M built for the Russian Navy, *Admiral Essen*. SDB

Length [m]	Autonomy [day]	Speed [knots]	Crew [man]	Displacement [t]	Width [m]
125	thirty	thirty	180	4000	15

Ship type: frigate
ADMIRAL MAKAROV

Conducting combat operations in the ocean and sea areas against surface ships and submarines of the enemy, as well as repelling attacks of air attack weapons both independently and as part of the formation of ships as an escort ship.

ARMAMENT

Artillery:
One single-barrel artillery installation A-190 caliber 100 mm
Two six-barreled artillery mounts AK-630 caliber 30 mm

Rockets:
Strike missile system "Caliber" (1 × 8 VPU)
The missile complex "Shtil-1" (1 × 24 VPU)

Anti-submarine:
Two two-tube torpedo tubes caliber 533 mm
1x12 RBU-6000

Top: The first Project 11356M Guard Ship built for the Russian Navy, *Admiral Grigorovich*, transits the Bosphorus as it departed the Black Sea on a cruise to the Mediterranean Sea on 4 August 2018. Bottom: Graphic depicting the Project 11356M Guard Ship, *Admiral Makarov*, with various particulars in reference to role, basic characteristics and armament. Press Service of the Southern Military District

Project 11356 Frigate (Guard Ship) Main Systems and Armament – data furnished by Severnoye Design Bureau

Fregat-M2EM radar: 1
MGK-335EM-03 sonar: 1
Vinyetka-EM sonar with flexible towed antenna: 1
Club-N missile complex with 8 missiles: 1
Shtil-1 surface to air missile complex: 1
A-190/E mounting for a single 100 mm calibre gun: 1
30 mm calibre automatic gun mounting: 2
DTA-53-11356 twin torpedo tube: 2
RBU-6000: 1
Ka-27/Ka-28 or Ka-31 helicopter: 1

Graphic depicting the Project 11356M Guard Ship (Frigate) armament and basic characteristics, which translates to an 8 cell vertical launch system for Kalibr surface to surface cruise missiles, a Shtil-1 surface to air missile with twenty four silo vertical launch system, a 100 mm calibre A-190 gun complex, two x AK-630M 30 mm calibre six barrel cannon systems, 533 mm calibre anti-submarine torpedo launchers, RBU-6000 anti-sub-surface rocket launcher, maximum speed of 30 knots, crew of 180 (Rosoboronexport documentation states 193) and displacement, 4000 tonne (full load weight is slightly in excess of 4000 tonne). MODRF

The Project 11356/M was designed for operation in a wide range of environmental climatic conditions ranging from Arctic to equatorial. Seaworthiness was improved over previous generation designs through incorporation of the UK6-1 roll stabilisation system (Rosoboronexport).

The project 11356/M Guard Ship/Frigate was designed with low signatures in the radio-electronic/electromagnetic, infrared and acoustic spectrums, facilitated through incorporation of signature reduction technologies and hull design. MODRF

The Project 11356/M design embraced traits/technologies to reduce the ships detectability in the radio-electronic/electromagnetic, infrared and acoustic spectrums (Rosoboronexport). To this end, the Project 11356/M ships were built with attention taken to reducing to the minimum, within the design constraints, the vessels acoustic, infrared and radio-electronic signature (Severnoye, 2013a). This was facilitated, in part, through incorporation of signature reduction technologies and hull design, which were instrumental in reducing the ships RCS (Radar Cross Section), electromagnetic, infrared and acoustic spectrums. The ship design and propulsion system were optimised for reduced underwater noise, in turn reducing the potential for noise radiating from the ship to interfere with the onboard sonar systems (Rosoboronexport).

The main propulsion system consisted of a single twin-shaft gas-turbine with a power output of ~28,000 hp. in each shaft, for a combined power output of ~56,000 hp., assuming an ambient temperature of +15° Centigrade. Operation of the propulsion system was controlled by a Burya-11356 control system. Electric power was provisioned through a quartet of 800 kW WCM800/5-type diesel fuel generators that fed the three-phase AC/380V/50Hz system, all controlled by an Angara-11356 complex (Rosoboronexport).

The first three domestic Russian Project 11356M ships were powered by gas turbines developed by Zorya-Mashproekt, Ukraine, three of which had been delivered before the Ukrainian coup of 2014 (Severnoye, 2016), which would end the delivery of such power units to Russia. By late 2020, Russia had solved the problem by developing domestic power plants for its Frigate type warship building program.

An Integrated Bridge Control System facilitated control of Project 11356/M ships and their respective technical elements during routine and 'special tasks' of onboard computer-aided C2 (Command & Control) systems when vessels so equipped were involved in operational tasks, such as tracking and or engaging a threat sub-surface target (Morinformsystem-Agat Concern).

The Project 11356/M design was equipped with the Trebovanie-M combat management/information and control system developed by NPO Meridian, a subsidiary of Concern Morinformsystem-Agat (USC & MODRF Press Service of the Western Military District, 2016a). Like the SIGMA-E, an alternative on the Project 11356, Trebovanie-M provided operational control of the host ship or task force through integrating radio electronic weapon systems into a single control system. The system embraced computer aided decision making in regard to employment of ship/task force weapons (Morinformsystem-Agat Concern).

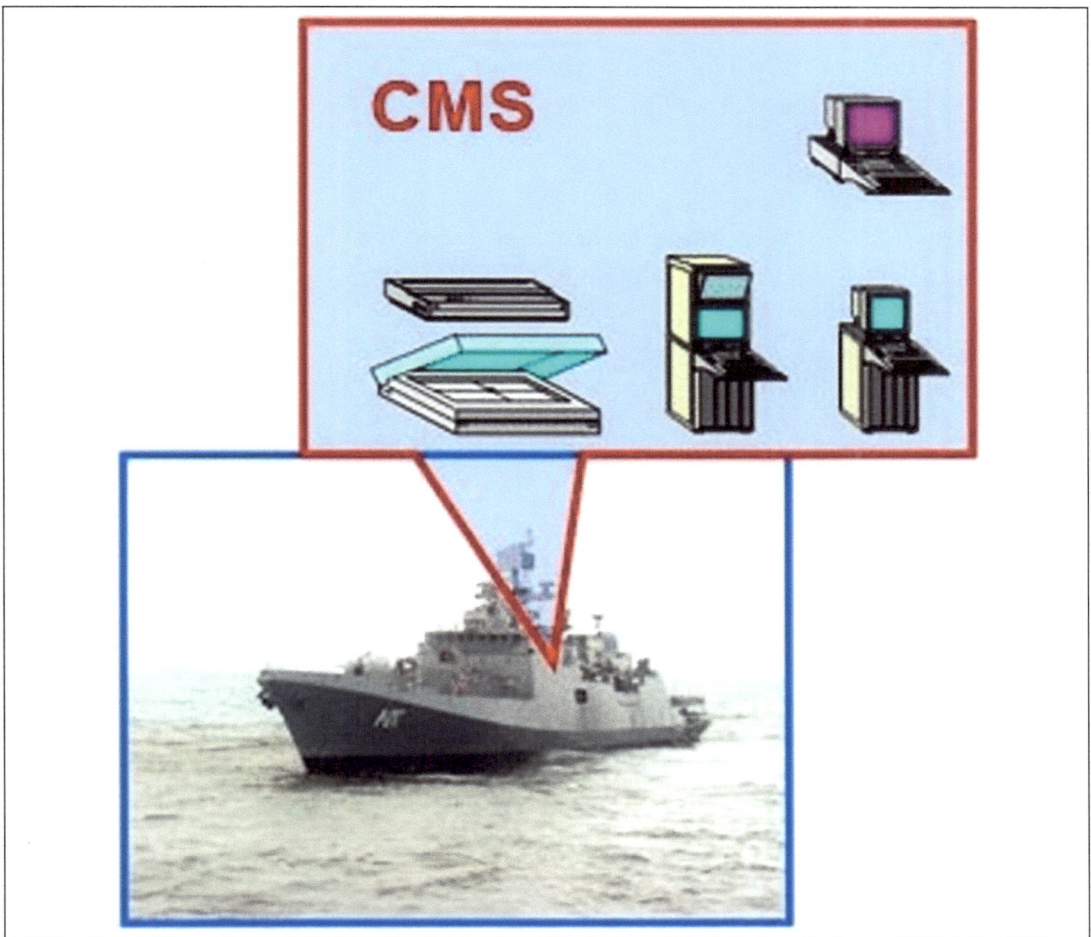

Diagram illustrating the presence of a Combat Management System/Combat Information and Control complex and main location of the major systems in the Project 11356/M Frigate. SDB

The Project 11356/M was equipped with a comprehensive navigation suite – MR-212/201-1 and Nucleaus-2-6000A navigation radars (Severnoye & USC) and the CSRI Elektropribor developed LADOGA-M/ME inertial navigation and stabilisation system for modern Russian submarines and surface warships. The inertial navigation and stabilisation system consisted of a unified gyro system, a digital computer, control unit and a heat setting system. The outflow of data from the system was available in analogue and digital MIL-STD-1553B data-bus form. Stabilisation parameters included 15 arc min for roll and pitch angles (not accounting for limiting errors), 0.2°/second for angular rates of roll, pitch and heading change and 0.1 (0.2) m/second for components of instantaneous motion velocity caused by roll, pitch and orbital motion at the place of the gyro device installation (Harkins, 2016 & CSRI Elektropribor, 2013).

LODOGA-ME navigational parameters – data furnished by CSRI Elektropribor

When using the data from a SNS receiver (adjustable mode)
Position (latitude, longitude) – along each coordinate: 0.4 km
Heading:
    3.0 arc min at latitude $\varphi \leq 60°$
    1.5 sec φ arc min at latitude $\varphi > 60°$
Northern and Eastern velocity components relative to the ground: 0.8 knots
When using data from shipboard log (autonomous mode)
Position (latitude, longitude) in period of 6 hours: 5 km
Heading:
    6.0 arc min at latitude $\varphi \leq 60°$
    3.0 sec φ arc min at latitude $\varphi \leq 6°$
Northern and Eastern velocity components relative to the ground: 1.2 knots
Power consumption: <1 kW

Diagrammatic view of a Project 11356 vessel indicating the presence of the Fregat-M2EM radar complex (inset). FSUE Salyut

FREGAT-M2EM – As briefly noted above, the Project 11356 design was equipped with a comprehensive radio-electronic (radar) suite for conducting airborne and surface surveillance and the detection and tracking of airborne and sub-surface targets. The primary radar sensor installed on the Project 11356 was the FSUE SMP Salyut developed Fregat-M2EM complex, designed for detection and tracking of airborne and surface targets. This radar complex emerged as an advanced development of the Fregat-MA that had been installed on previous generation Russian warships (FSUE Salyut, Severnoye and Harkins, 2017). The Fregat-M2EM was developed as a 3-D (3-Dimensional), two-channel noise-proof radar complex optimised for detection of airborne and sea surface targets and for processing targeting data for designating targets for engagement by shipborne weapons. The major functions of the radar complex included: providing surveillance of the air and sea surface environments at ranges out to 300 km and altitudes up to 30 km (this relates to air); detection of airborne targets, including small size low signature objects, at a range of altitudes from medium/high to low; detection of large and small surface targets; initial primary radar data allocation to ship weapon systems; providing integral ECM (Electronic Countermeasures) protection for the data processing system; provision for transfer and processing target information to processing station(s); allocation of secondary radar data to relevant control stations and transfer of target allocation and designation information for ships defensive weapons (FSUE Salyut).

Fregat-M2EM two-channel noise-proof radar complex. FSUE Salyut

Project 11356M Guard Ships *Admiral Essen* (top) and *Admiral Makarov* (bottom), illustrating the prominent position of the Fregat-M2EM radar antenna on the Project 11356/M vessels. MODRF

The Fregat-M2EM features high levels of anti-interference (jamming) through incorporation of a number of functions: 'Adaptive distribution of energy in space'; Wide-band electronic retuning; 'High space resolution and low level of antenna pattern sidelobes'; handling of complex radio-electronic signals in high resolution; Nonlinear processing of received signals; blocking out distorting echo-signals by the antenna sidelobes from an object featuring significant surface for reflecting radar returns; Objective post detection processing of signals and 'double-adaptive moving target selection' (FSUE Salyut). The radar complex featured the ability to automatically conduct diagnostics to counter faults (FSUE Salyut).

Fregat-M2EM technical characteristics – data furnished by FSUE SMP Salyut

Frequency band: E
Number of radar channels: 2
Number of measured coordinates: 3
Range: 300 km
Azimuth: 360°
Altitude: 30 km
Elevation
    I channel: 45°
    II channel 55°
Detection range against airborne targets: 230 km for a tactical combat aircraft size target or 50 km for a cruise missile size target
Detection range against sea surface targets: range of direct visibility
Minimum range: 2 km
Maximum scan rate: 2.5 seconds
Antenna rotation speed: 12.6 revolutions per minute
Measuring accuracy of coordinates
    Range: 120 km
    Azimuth: 5 mils
    Elevation: 7 mils
Weight of antenna post: 2.5 tonne
Weight of devices: 9.25-9.85 tonne (dependent on composition)
Power consumption: 90 kW
Number of devices: 20-24 (dependent on composition)

3Ts-25/E – The Project 11356/M was equipped with a 3Ts-25/E target acquisition/designation radar complex, which consisted of active and passive radar channels and connected multiband antenna and data processing systems. The 3Ts-25E employed active and passive channels to provision for the surveillance of the sea surface situation and objects upon it in order that they could be designated for attack with anti-ship missiles. The functions of the complex included: detection of, classification of and determination of the surface target coordination through analysing of target radar emissions by the 3Ts-25E passive channels; employment of integral interrogator to determine target origin (nationality) through identification friend or foe) interrogation; data would be transferred from the ship 3Ts-25E to the anti-ship missile weapon control system receivers and the ships combat data management system; the 3Ts-25E could communicate with external platforms, such as others ships in a task grouping or airborne platforms and provisions for tactical level navigation and maneuvering to increase ship safety by avoiding hazards (Rosoboronexport).

The antenna and main complex for the 3Ts-25/E target acquisition/designation radar system was located on the deck atop the bridge section of the Project 11356/M as shown on the *Admiral Grigorovich*, **carrying side code 494 (top) and** *Admiral Grigorovich* **carrying side code 745 during sea trials (bottom).** Rosoboronexport/NPO Meridian

3Ts-25E in opposing segments of the antenna rotation. Rosoboronexport

The long-range capability, when operating in dense ECM and rough sea surface environments, is facilitated through incorporation of high power output and modulation of signals integrated with the systems passive channel. The passive channel analyses and classifies intercepted radar signals (emissions) through a comparison of signals from a comprehensive (around 1,000 inputs) database. The active channel, which, through a multi-computer system, is integrated into secondary data processing and the wider network, sends out probing signals through use of a multiprocessor system and data from the radar complex (Rosoboronexport).

3Ts-25E radar complex – data furnished by Rosoboronexport

ACTIVE CHANNEL
Frequency band: I (this corresponds to the band utilised by NATO and is accurate for export Project 11356 Frigates, but it is unclear if this corresponds to the domestic Russian Project 11356M fleet)
Probing signal types: Pulse, complex with intrapulse phase manipulation
Scan modes: Sector and circular
Range of target detection (assuming a RCS of 1000 m^2 – dependent upon area to be covered): 35-45 km in conditions of normal radar visibility, up to 90 km in conditions of optimum (high) radar visibility and up to 250 km with super-refraction resolution when operating in active mode (dependent on range)
Range: 40-960 m
Elevation: 1.5°
PASSIVE CHANNEL
Sensed signal at frequency band (complete waveband coverage): centimetric-wave, decametric-wave
Sensed signal types: pulsed, continuous with random polarisation
Surface target detection range: 50-500 km (dependent on signal potential and frequency band of radar emissions)
Root-mean square range error: 4-20%
Root-mean square bearing error: 0.5-2°
Target classification: probabilistic classification sensed coastal radars using customer war models

---

As briefly noted above, the Project 11356/M was equipped with a comprehensive sonar suite for the detection of aquatic objects, such as submarines, sea surface or sub-surface drifting hazards. Severnoye documentation indicates that the Indian Teg Class is equipped with a DRDO developed and built M/S Bharat Electronics HUMSA (Hull Mounted Sonar Advanced) (Severnoye). In the Russian domestic Project 11356M, the primary sonar complex is the MGK-335EM-03, which was developed for installation in surface vessels of small to medium size/displacement (Concern Okeanpribor). The major functions of the MGK-335EM-03 include detection of underwater objects, such as a submerged submarine, whilst operating in active mode; automatic target tracking; transfer of targeting data for designation of specific target(s); location of underwater objects whilst operating in passive mode through detection of noise emissions emanating from the target; conduct low-frequency and high-frequency hydro-acoustic communications, 'code communications and recognition'; detection of radiation signals from potential target automatic classification of located target(s); reduced acoustic interference and automated diagnostic detection of faults (Concern Okeanpribor).

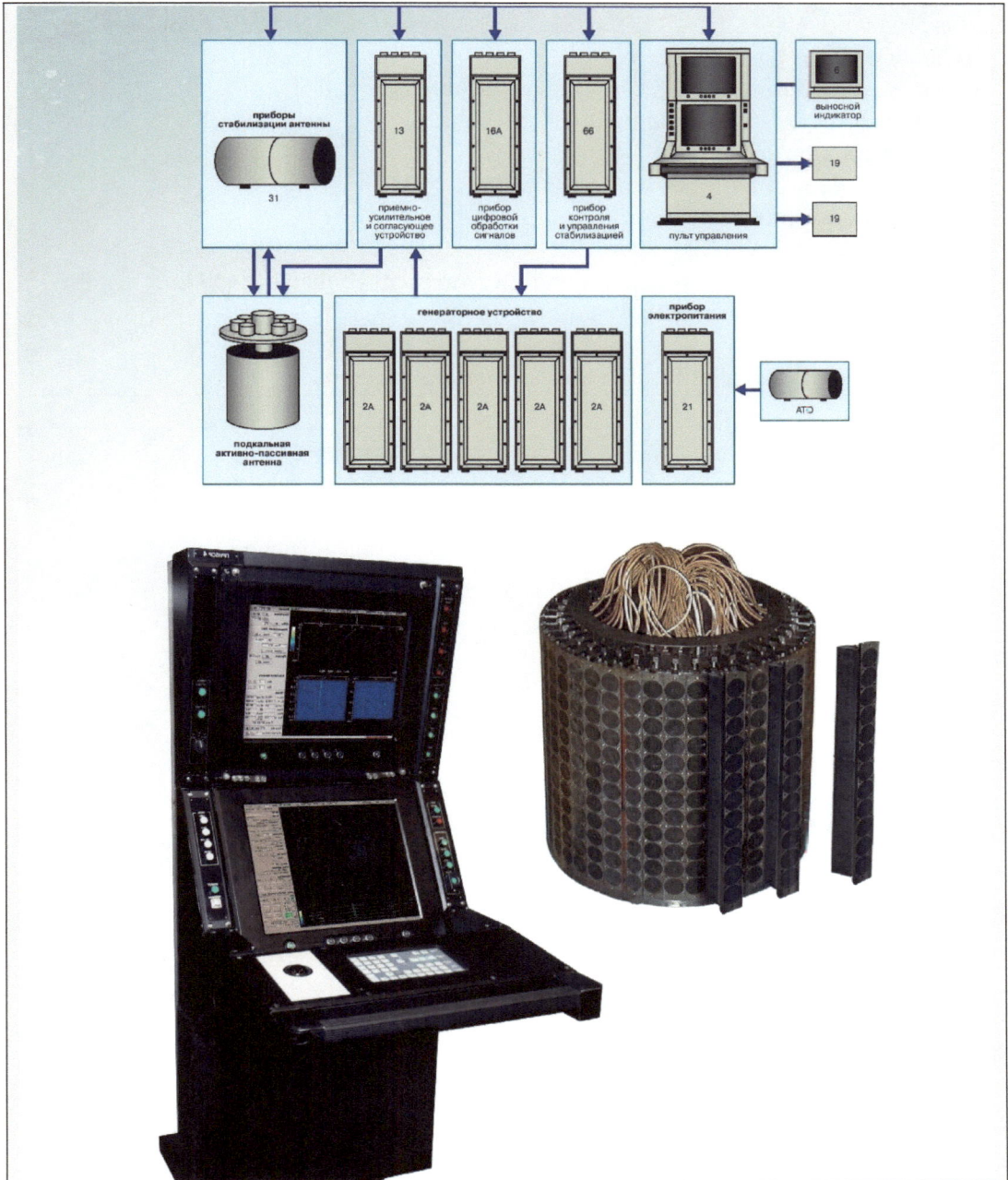

MGK-335EM-03 complex and control system. Concern Okeanpribor

MGK-335EM-03 technical characteristics – data furnished by Concern Okeanpribor

Sonar detection range: Typically 10-12 km
Determination of target coordination accuracy
    Bearing: 0.5°
    Range: 1% of the distance scale
Noise reduction
Frequency range: 1.5-7 kHz
Accuracy of determining target bearing in tracking mode: 0.5°
Detection of hydro-acoustic signals (OGS)
Frequency range: 1.5-7 kHz
Accuracy of determining bearing to the source: 10°
Hydro-acoustic communication
Energy range: TTS NH VH: 20 km
Code communication, identification and measurement of distance: 30 km
Telephony frequency range: Present
Classification
Operator automated classification: anti-submarine torpedo

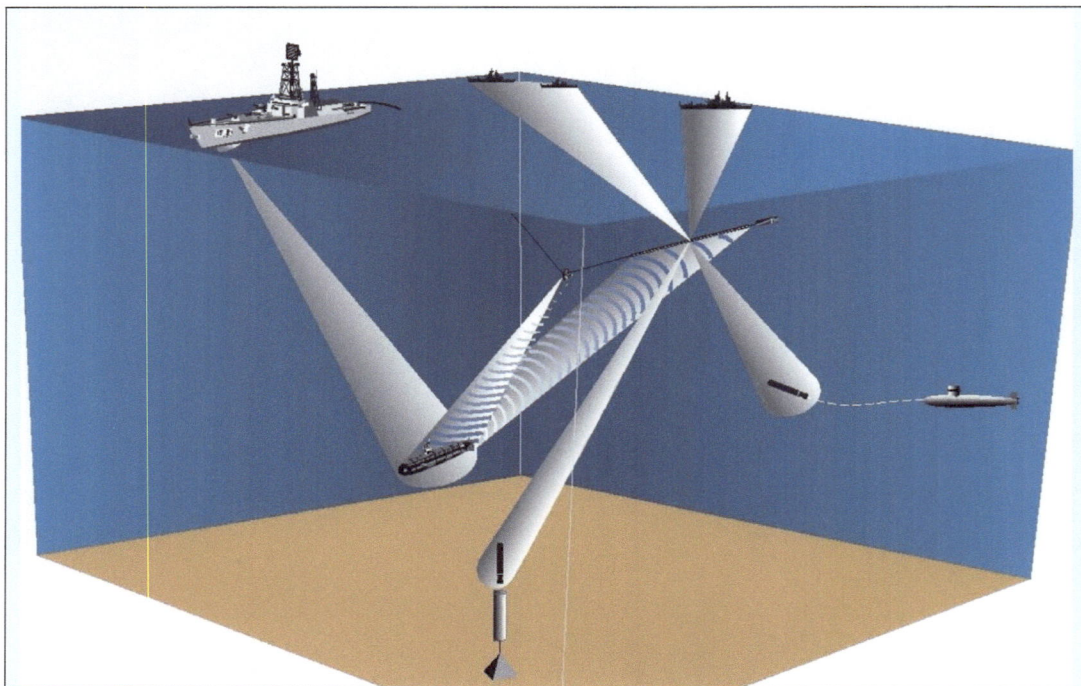

Graphic depicting types of targets that could be detected and classified by the Vinyetka-EM sonar, with flexible towed antenna complex – submerged submarines, sea mines anchored to the sea floor and torpedoes launched by submarines and surface vessels. Concern Okeanpribor

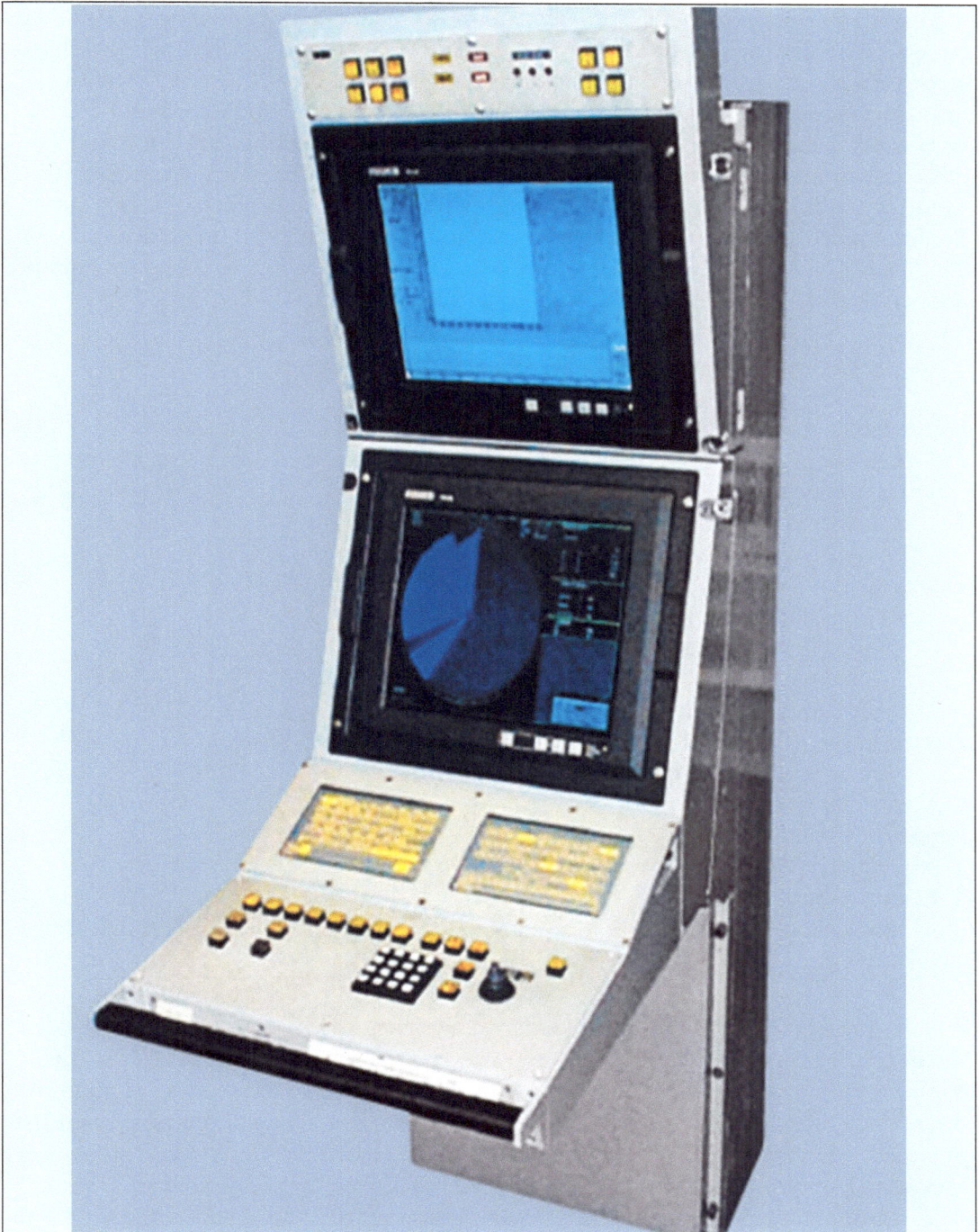

Console for the operation and control of the Vinyetka-EM sonar, with flexible towed antenna. Rosoboronexport

Vinyetka-EM – The Vinyetka-EM sonar, with flexible towed antenna, was designed and developed as a low-frequency active-passive hydro-acoustic complex for installation on surface ships to facilitate the detection and tracking of sub-surface targets. The complex included a towed antenna (trailed astern of the ship), incorporating a low-frequency stationary emitter tasked with active detection of modern low-noise emitting submarines. The system was primarily intended for the detection of sub-surface targets in sonar mode and noise direction location. The noise direction finding mode allowed for the detection of long-range torpedo threats to the host ship and surface vessels at long-range. Bearings of detected targets would be determined and the target tracked with automated classification (Okeanpribor). The SNN-137 sonar with towed active array was specified as a military technical cooperation alternative to the Vinyetka-EM (Severnoye).

Vinyetka-EM sonar – data furnished by Concern Okeanpribor

Sonar detection range
    In shallow water: Typically 10-20 km for a submarine (dependent on varying target parameters)
    In deep water: Typically 10-60 km
    Sector Review: $\pm180°$
Determination of target coordination accuracy in automatic tracking mode
    Bearing (in traverse corners and on straight track): 2°
    Range: 1% of the distance scale
Towing speed: Up to 18 knots
Noise reduction
Detection range against an actively maneuvering submarine: 15-20 km
Detection range against surface vessels: 30-100 km
Detection range against torpedo target(s): 15-30 km
Direction finding accuracy: 2°
Towing speed: Up to 14 knots
Target classification
Operator automated classification of targets for anti-submarine torpedo or surface attack

---

For protection against threat radar complexes and radio-electronic guided weapons, the Project 11356/M vessels were equipped with an ASOR-11356 ECM complex tasked with interrupting the detection, tracking and targeting capabilities of threat weapons (Severnoye & USC). A PK-10 short-range jamming complex provided the host ship platform with short-range multi-class decoy protection against radar and optical guided anti-ship missiles and other non-anti-ship optimised radar/optical guided weapons. The complex consisted of a KT-216-E launch system – this could be 2, 4, 8, 12 or 16 launchers (two launchers installed on the Project 11356/M) – for three types of projectile, A3-SO-50, A3-SOM-50 and A3-SR-50. The first two types could be used against optical guided weapons and the third could be employed against radar guided weapons. The decoy projectiles would be launched to a point in close proximity to the host ship with the intention of decoying threat missiles away from homing on the ship by producing multiple targets with signatures more prominent that that of the missiles intended target – the ship. This would be accomplished through either causing the missile homing head to break lock/track completely or lock onto a decoy mimicking a more appealing target (Rosoboronexport).

Top: Towed antenna array of the Vinyetka-EM sonar complex. Bottom: Lead Project 11356M ship, *Admiral Grigorovich*, transiting to Sevastopol, Crimea. Concern Okeanpribor/MODRF

Decoys could be launched from the KT-216-E system day or night in environmental conditions of fair or adverse weather. Launcher loading would be conducted manually with typical composition being SO rounds or SR rounds in one launcher. Operation of the PK-10 complex could be conducted through a remote control unit, the status of individual rounds, SO or SR, being monitored through the primary automatic control system or remote control unit (Rosoboronexport).

Ten tube PK-10 short-range jamming complex for multi-class decoys for protection against radar and optical guided precision guided weapons. Rosoboronexport

PK-10 projectile characteristics – data furnished by Rosoboronexport

Round type: A3-SR-50
Calibre: 120 mm
Round weight: 25.5 kg
Filler weight: 10 kg
Filler type: Chaff
Length: 1226 mm
Operating temperature range: -40° C to +50° C

Round type: A3-SO-50
Calibre: 120 mm
Round weight: 20 kg
Filler weight: 6 kg
Filler type: Infrared, laser
Length: 1226 mm
Operating temperature range: -40° C to +50° C

Round type: A3-SOM-50
Calibre: 120 mm
Round weight: 21.8 kg
Filler weight: 7.3 kg
Filler type: Infrared, laser
Length: 1226 mm
Operating temperature range: -40° C to +50° C

Not confirmed, it is plausible that the Project 11356M was, or may at a future modernisation, be equipped with an optical-electronic deflection system, such as the JSC Typhoon developed MDM-2/E. This system was designed to protect the host platform from semi-active laser homing precision guided munitions.

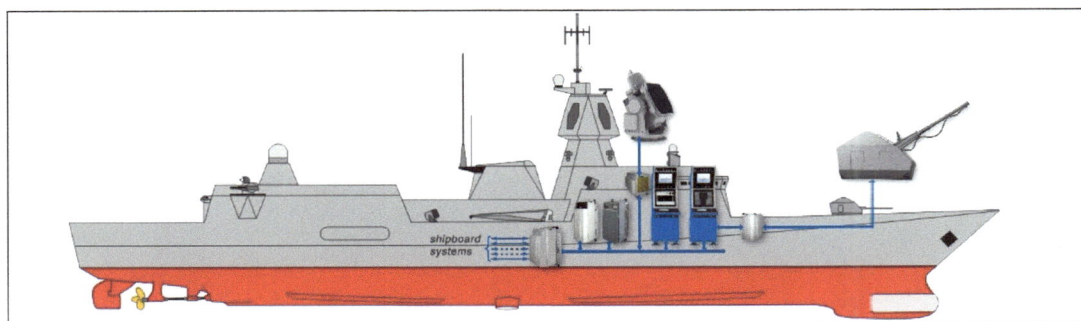

Graphic depicting the constituent elements of the combined A-190-01/A-190E-5P-10E gun/fire control radar complex and integration with shipboard systems of a notional warship design. KB Ametist

The A-190-01 is the primary gun armament on a number of Russian warship designs, including the Project 20380, 20386, 21630, 21631 and 11356. Rosoboronexport

A-190-01/A-190E-5P-10E – The primary gun armament of the Project 11356/M consisted of a single gun A-190-01 universal lightweight artillery turret complex (domestic Russian ships) or A-190E (Indian ships) designed to function in a dense jamming environment. The gun is controlled by a 5P-10/E Puma fire control radar system, making up the combined A-190/E-5P-10/E complex. A magazine, located in the lower compartment of the lightweight universal artillery mounting, could accommodate 80 rounds of 100 mm ammunition (Rosoboronexport & Severnoye) – there are two values forwarded as total 100 mm ammunition capacity accommodated in the Project 11356/M – 500 shells (Rosoboronexport) and 300 shells (Severnoye).

The A-190-01 could fire unitary loading high explosive fragmentation rounds with shock fuse for use against surface targets or remote fuse for use against airborne targets (CRI Burevestnik). Gun loading was accomplished through an independent ammunition feed system, taking rounds from port and starboard magazines. This provided timely availability of two separate types of ammunition, one from each magazine, which contributed to a high rate of fire in comparison to legacy systems. The gun complex could be used in a quick reaction automated mode against a number of surface and airborne target sets. Surface targets that could be accurately engaged at long-range varied in composition from fixed coastal targets to mobile and maneuvering sea surface targets (Rosoboronexport). The artillery complex could be used at long-range against multiple surface and airborne targets, with a reaction time of 2-5 seconds against air targets, with the ability to quickly switch from one target to another in a multi-target environment (CRI Burevestnik).

The A-190-01 gun turret was positioned near the Project 11356M bow section, forward of the vertical launch complexes for the Shtil-1 and Kalibr-NK missiles. MODRF

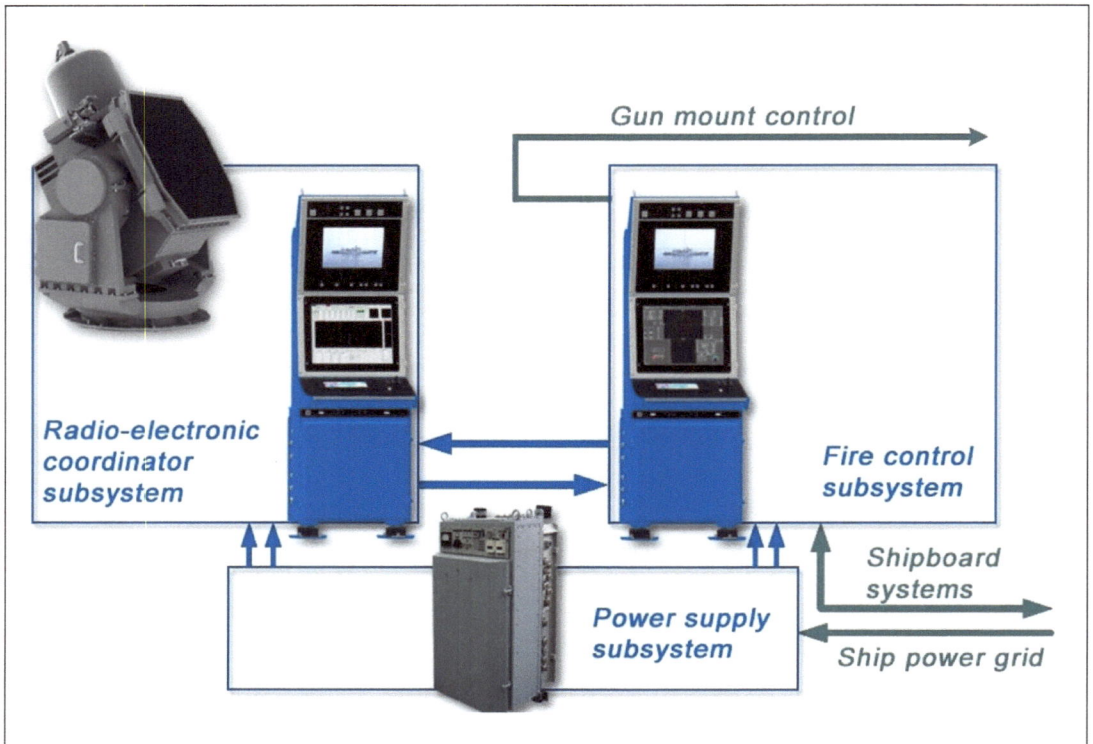

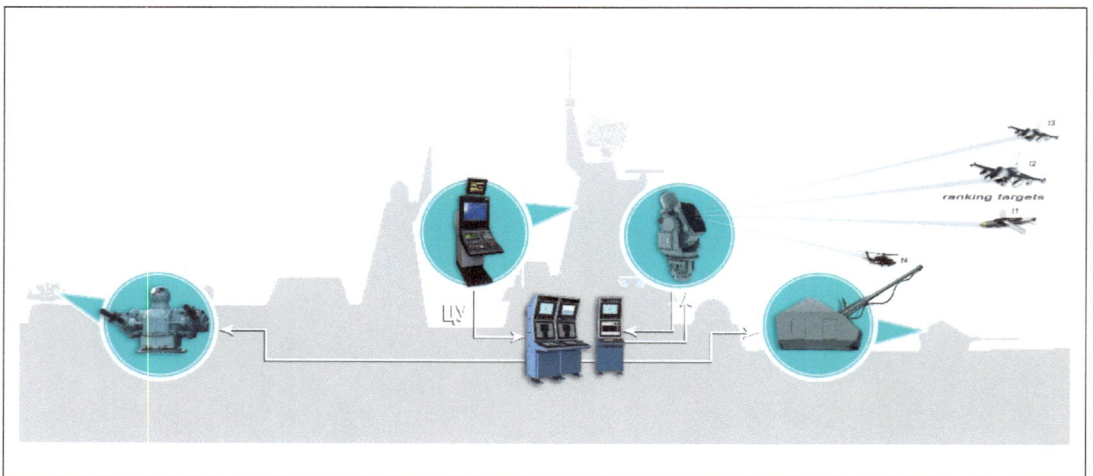

Top: Graphic depicting the various components of the 5P-10E and the process of data transfer between the complex, the gun systems, shipboard systems, integral power supply and shipboard power grid. Bottom: Graphic depicting the 5P-10E control process for targeting various airborne target types by the shipboard primary and secondary gun systems (notional ship design). KB Ametist

The 5P-10E could provide all-round and sector search for targets autonomously, and prioritise targets according to threat analysis. Lock-on to target would be conducted automatically, with up to four targets simultaneously within the scanning field of the antenna. Target data would be automatically received by the ship control systems and coordinates and parameters of target motion passed on to the ships targeting distribution system for potential engagement by onboard weapon systems. This allowed the target, be it airborne, seaborne or coastal, to be simultaneously engaged by multiple weapon systems, such as the A-190 and the 30 mm automatic cannon complexes. The 5P-10E provisioned for calculating miss-distance automatically, allowing fire to be adjusted onto the target. The complex also provided for crew training against simulated targets and fault diagnostics in automated mode, providing recommendations on the best solution for correcting diagnosed faults (Rosoboronexport).

As well as controlling the operation of the A-190/E, the 5P-10E had the function of testing the complex and provided automatic ammunition type selection. In the event of a complete power failure the targeting function of the A-190/E could be conducted manually through employment of an optical sight system (Rosoboronexport).

A-190-01/A-190E-5P-10E – data furnished by CRI Burevestnik, Rosoboronexport, KB Arsenal & Severnoye Design Bureau

Rate of fire: ≤ (less than or equal to) 80 rounds per minute
Range of fire, lateral: more than 20000 m
Elevation angles: -15° to +85° and ±17°, horizontal
Projectile weight: ~15.6 kg
Ammunition capacity of A-190 mounting: 80 rounds (300 rounds ship capacity – Rosoboronexport states 500)
Mounting weight: ≤15000 kg
Data acquisition channels: Radar and electro-optical
Radar frequency band: 3 cm
Fire control system limits in all-round surveillance mode
    Relative bearing: 360°
    Elevation angle: -5° to +40°
    Range: ≤30 m
Fire control system limits in tracking mode
    Relative bearing: ±200°
    Elevation angle: -5° to +85°
    Range: ≤60 m
Scan period: 1.5-10 seconds
Combat readiness alert time from cold status: ≤3 minutes
Response time for firing against a target: 5 seconds for an airborne target and 10 seconds for a sea surface target
Fire shift time within 6 x 3 tracking sector: 1 second

The Project 11356/M was armed with a secondary gun armament in the shape of two AK-630M six barrel AO-18 automatic gun CIWS for employment against airborne and surface targets. The AK-630M modules were positioned on the port and starboard flanks of the helicopter hanger complex in the aft section of the ship, the starboard module shown on the *Admiral Grigorovich* on her transit to Sevastopol from the Mediterranean Sea. MODRF

The AK-630/M 30 mm six barrel cannon modules were developed for employment against targets that had penetrated the outer layers of air defence. Developed as a CIWS (Close In Weapon System) to supersede the AK-230 twin 30 mm cannon mount, the AK-630/M would be capable of engaging high speed targets, such as an ASCM (Anti-Ship Cruise Missile), that had penetrated the outer air defence layers of a warship or a group of warships. The AK-630/M could also be employed against other airborne target sets, ranging from slow moving helicopters to subsonic and supersonic aircraft. The weapon could also be employed against surface targets, including small warships and other waterborne objects, including drifting mines, and against shore targets.

The AK-630M integral AO-18 automatic gun system featured a continuously cooled six-barrel 30 mm gun, fed ammunition from the magazine (capacity of 2,000 rounds) through an automatic belt feed system. The breach-block mechanism would ram the round home and eject the casing after expenditure. The gun, mounted in a revolving turret, was a high velocity water cooled cluster system able to lay down a barrage of 30 mm projectiles into the path of an oncoming target through remote control gun laying, courtesy of the fire control system and the sighting station.

AK-630 combat module with six barrel AO-18 automatic gun system. Tulamashzavod

AK-630M Automatic Gun Mount – data furnished by Concern-Agat

Calibre: 30 mm
Maximum engagement range: 5000 m
Rate of fire: Up to 5,000 rounds per minute
Muzzle velocity: 875 m/second
Laying angles: -12° to +88° in elevation and ±80° in traverse
Ammunition load: 2,000 rounds in magazine
Gun mount weight (without ammunition): ~1000 kg

The composition of the Shtil-1 multi-channel medium range surface to air missile complex was different in the Indian Project 11356, Talwar & Teg Class and the Project 11356M for the Russian Navy. The Indian Navy ships were equipped with a single-rail launch system abaft the A-190E gun module (top) whilst the Russian Navy ships were equipped with a below deck 3S90 vertical launch system for the rapid fire of missiles (*Admiral Essen* illustrated bottom). USC/Rosoboronexport

Shtil-1 multi-channel single-rail launch medium range surface to air missile system.
Almaz-Antey

SHTIL-1 – The Project 11356 primary air warfare system was the Shtil-1 multi-channel medium range surface to air missile complex with a complement of 24 missiles. Shtil-1 is an omni-directional air defence complex designed to provide the host platform with the ability to engage a plethora of airborne target types in defence of the ship or in defence of other vessels in a task force grouping and ships under escort. Shtil-1 can be employed against high performance multi-Mach aircraft, slow moving helicopters and drones (Uninhabited Air Vehicle), anti-ship cruise missiles or sea surface targets – ships and fast attack craft/patrol craft etc. (Severnoye, Almaz-Antey & Rosoboronexport).

The Shtil-1 complex was composed of a number of systems – a 3R90E1 multichannel fire control system; a 3S90/E honeycomb effect vertical launch system for the rapid fire of missiles (domestic Russian fleet ships) or a single-rail launch system (Indian ships) and 9M317E (rail-launch) or 9M317M/E (vertical launch) missiles; pre-launch preparation system (modular design); software system; support system and system for basic maintenance of the complex (Rosoboronexport).

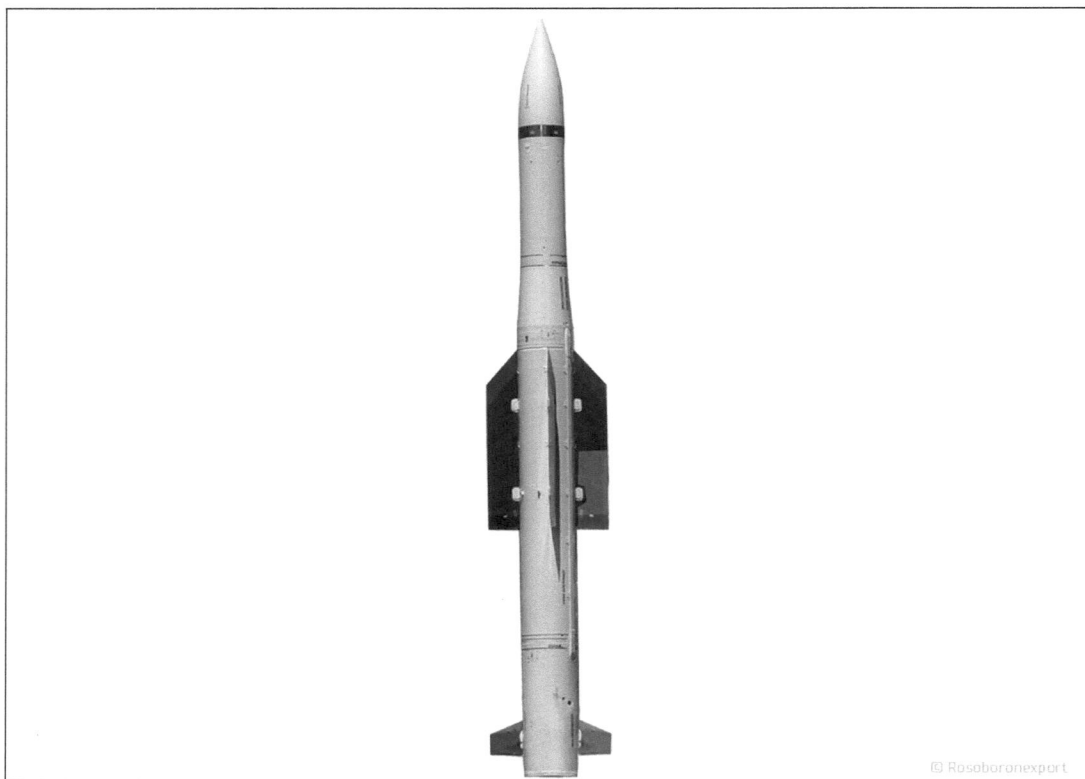

Depending on complex type, the Shtil-1 could engage up to 12 targets simultaneously (the single-rail launch variant could engage two targets simultaneously). Standard procedure would be to launch between one and three missiles at each target. The determining factors on the number of targeting channels and the stock number of missile/launch silos is primarily dependent upon ship displacement (the ability to accommodate the launchers/missiles) and the requirements of the intended operator. In domestic Russian Federation service the Project 11356M is equipped with 24 ready to launch missiles housed in 24 vertical launch silos (Rosoboronexport).

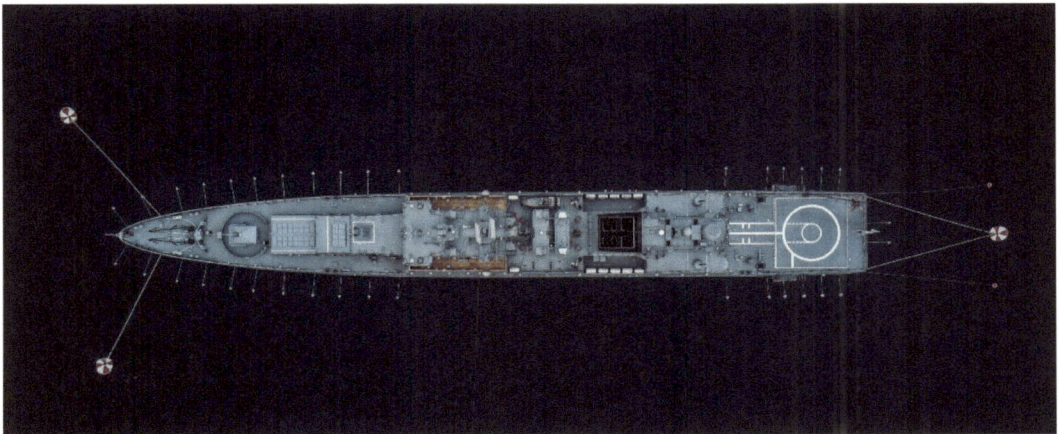

Previous page: 9M317M/E vertical launch surface to air missile. This page 3S90/E honeycomb effect vertical launch diagram for 36 missiles (complex installed on the Project 11356M accommodates 24 missiles) (top) and Project 11356M plan view showing 3S90 complex aft of the A-190-01 gun complex (bottom). Rosoboronexport

The missiles were designed as single-stage solid state vehicles. The targeting, launch and engagement process would be conducted in automatic mode. Targeting information was transferred to the missile from the ships 3-D all round surveillance radar complex. After launch the missile would be guided to the target through inertial navigation + radio correction (9M317ME1) with semi-active radar homing in the terminal phase of the flight to the target. The warhead would be detonated through impact or radar proximity fuse. As well as controlling operation of the Shtil-1 complex, the radar could be used to control the ship artillery systems, primarily the A-190-01/E complex (Rosoboronexport).

5P-10-03E Universal small size fire control system antenna. Rosoboronexport

Top: A 9M317E surface to air missile is launched from a Shtil-1 multi-channel single-rail launch medium range surface to air missile system of a warship during weapon trials. Bottom: INS *Talwar* (F40) with Shtil-1 single rail launcher positioned aft of the A-190E gun complex. Almaz-Antey/SDB

Shtil-1 characteristics for single-rail launch – data furnished by Rosoboronexport and Almaz-Antey

Launch range parameters: 3.5 to 32 km
Maximum slant range (altitude): Up to 15000 m
Minimum slant range: 5 m
Maximum speed of targets that can be engaged: 830 m/second (~2988 km/h)
Operational mode (main): Automatic
Number of targets that can be simultaneously engaged: 2-12 (dependent upon system composition)
Deployment time from stand by: 10-19 seconds
Rate of fire (one launch complex): 12 seconds
Missiles per launcher: 24
Number of launchers in complex: Up to four (Project 11356 ships are armed with one launch complex)
Maintenance crew: 11-32 dependent upon composition
Magazine operating environment parameters: -5° to +35° Centigrade
Humidity parameters: +30° Centigrade ±5° Centigrade
Maximum wind speed operating parameters: 20 m/second (72 km/h)
9M317E missile
    Length: 5.55 m
    Diameter: 0.4 m
    Wing span: 0.86 m
    Launch weight: 715 kg
    Maximum flight speed: Mach 3
    Guidance: Inertial navigation with radar homing in terminal phase
    Warhead type: High explosive-fragmentation
    Warhead weight: 70 kg

Project 11356M Guard Ship *Admiral Grigorovich* **of the Russian Black Sea Fleet.** MODRF

Shtil-1 characteristics for vertical launch system – data furnished by Rosoboronexport

Launch range parameters: 3.5 to 50 km
Maximum slant range (altitude): Up to 15000 m
Minimum slant range: 5 m
Maximum speed of targets that can be engaged: 830 m/second (~2988 km/h)
Operational mode (main): Automatic
Number of targets that can be simultaneously engaged: 2-12 (dependent upon system composition)
Deployment time from stand by: 5-10 seconds
Rate of fire (one launch complex): 12 seconds
Missiles per launcher: 12
Number of launchers in complex: Up to four – Domestic Russian Project 11356M ships are armed with two 12 missile vertical launch systems
Maintenance crew: 6-27 dependent upon composition
9M317ME1 missile
    Length: 5.18 m
    Diameter: 0.36 m
    Wing span: 0.82 m
    Launch weight: 581 kg
    Maximum flight speed: Mach 4
    Guidance: Inertial navigation plus radio correction with radar homing in terminal phase
    Warhead type: High explosive-fragmentation
    Warhead weight: 62 kg

For engagements against sub-surface targets the Project 11356/M was armed with a single RBU-6000 anti-submarine rocket projector and two DTA-53-11356 twin torpedo tubes for launching 4 x SET-65E/53-65KE anti-submarine warfare torpedoes. Anti-submarine weapons employment would be controlled by the Purga-11356 fire control system through data transferred from ship sonar (Severnoye).

RBU-6000 (Reactive Bombometnaya) – The RBU-6000 12-barrel rocket/missile launcher fired 90R anti-submarine homing projectiles or RGB-60 depth bombs. These, together with a fire control system, ammunition loading system and various support infrastructure, constituted the RPK-8 complex (Rosoboronexport). In the Project 11356/M the RBU-6000 armament consisted of 48 x 90R/RGB-60 depth bombs) (Severnoye).

The ship sonar systems – MGK-335EM-03 or Vinyetka-EM sonar with flexible towed antenna – provided targeting data for designation of RBU-6000 projectiles (Rosoboronexport). The projectiles could be rocket propelled to a maximum range of 4300 m (minimum launch range against underwater targets was 600 m) from the launch platform. The gravitational submerging section would separate from the 90R boost section, enter the water and home on the target (JSC Scientific Production Association Alloy & Harkins, 2017AC2).

The Project 11356/M was armed with a single RBU-6000 12-barrel rocket/missile launcher for 90R anti-submarine projectiles or RGB-60 depth bombs. The launcher, projectile loading system, projectiles, fire control system and support infrastructure constituted the RPK-8 anti-sub-surface complex. Rosoboronexport

Starboard side-on (top) and front hemisphere (bottom) views of an RBU-6000 complex. Rosoboronexport

RBU-6000 (Reactive Bombometnaya) basic specification – data furnished by JSC Scientific Production Association Alloy & Rosoboronexport

Caliber: 212 mm
Number of launch tubes: 12
Weight of entire complex, without ammunition: 9000 kg
Weight of the launcher, unloaded: 3500 kg
Weight of ammunition loader/feed system: 4700 kg
Above deck complex dimensions: 2600 x 2140 mm
Below deck dimensions: 760 x 1300 mm
Maximum elevation provided by electric power drive: 60°
Maximum descent angle: -90°
Angle of vertical guidance for firing at maximum range: 46°
Angle of vertical guidance at minimum firing range: 8.5°
Maximum firing arc: 340°
Guidance speed
In all rows: 27°/second
On the horizon: 27°/second
Ready to fire time from target detection: 15 seconds maximum

The Project 11356M Guard Ship *Admiral Essen* during weapon firing (RBU-6000 in foreground) at ranges in the Black Sea, circa 2019. MODRF

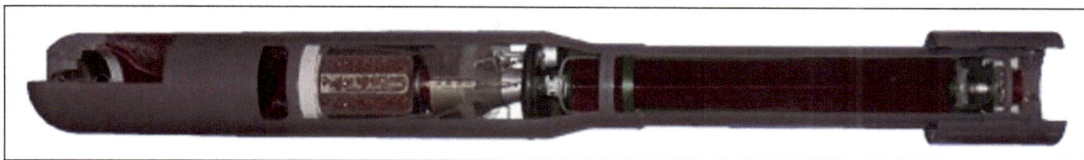

90R missile of the RBU-6000 complex. JSC Alloy

RBU-6000 complex projectile specifications – data furnished by JSC Scientific Production Association Alloy, KB Arsenal & Rosoboronexport

90R missile weight: 112.5 kg
RGB-60 bomb weight: 113.6 kg
Projectile length: 1832 mm
Warhead weight: 19.5 kg
Firing range of missile 90R & RGB-60
Maximum: 4300 m
Minimum: 600 m (Rosoboronexport states 210 m)
Engagement depth: 1000 m against submarine target and 4.10 m against torpedo and saboteurs (value relates to 90R)
Claimed salvo hit probability: 0.8 of 1

Russian Navy Project 11356M ships may be armed with the upgraded Shell Sonar Interference MG-94ME for launch from the RBU-6000. This projectile is designed to interfere with the guidance of a threat torpedo launched at the host ship by means of a broadband barrage – sighting frequency with a range of emitted frequencies that correspond to the various frequencies of target torpedoes (JSC Scientific Production Association Alloy).

SET-65E/53-65KE – The SET-65E/53-65KE anti-submarine warfare torpedo was developed as a 533 mm calibre class electrically driven homing weapon. The torpedo, which could be launched from submarines as well as surface warships, was stored and transported inside a watertight nitrogen filled container. The main sections of the torpedo consisted of an electric propulsion unit, featuring a disposable self-activated battery, the control unit, the high explosive warhead, an ECM-resistant active/passive homing system and proximity and contact fuses. The stand-out performance traits of the weapon were its high speed, long-range and fire and forget capability, courtesy of the autonomous functionality built into the system. The homing/guidance system, in conjunction with the yaw/depth/heel control system, which allowed the weapon to conduct two-plane manoeuvres, would guide the torpedo to the target, which would be destroyed or disabled by the proximity or impact exploders. The torpedo was classed as wakeless, with a constant speed and engagement range regardless of what depth the weapon was at within its operating parameters.

Technical Specification SET-65KE – data furnished by Rosoboronexport

Calibre: 533 mm class
Length: 7728 mm
Weight
    Live torpedo: 1703 kg
    Practice torpedo: 1342 kg
Range: Up to 16000 m
Speed: 40 knots
Target detection range: Up to 1500 m
Submarine engagement depths: 27-400 m
Homing: active/passive sonar with phased direction finding

SET-65KE homing torpedo. Rosoboronexport

CLUB-N/KALIBR-NK – The Project 11356/M was equipped with a 3S-14/E underdeck vertical launch complex fitted with eight launch tubes, primarily for 3M-54TE anti-ship missiles, contained in integral TLC (Transport Launch Containers). The missile/launch complex, with a missile loading system, a 3R-14N-11356 fire control system and 3C-25E target acquisition radar, were an integral element of the Club-N complex (Severnoye & Rosoboronexport).

Club-N was designed as a surface launched integrated anti-ship/anti-submarine/land attack missile complex designed to operate against respective targets in the face of intensive active and passive countermeasures. Club-N effectively referred to what was described as a unified combat system comprising two separate types of anti-ship cruise missile and an anti-submarine ballistic missile developed by JSC Experimental Machine Design Bureau Novator. The 3M-54E anti-ship cruise missile was composed of several sections – a booster, a low-altitude subsonic sustainer section and a separable supersonic warhead section (200 kg), whilst the 3M-54TE added an integral TLC for vertical launch. The 3M-54E1 consisted of a booster stage and a low-altitude subsonic sustainer, the 3M-54TE1 adding the TLC for vertical launch – these variants were equipped with a 400 kg warhead. The 91RE2 (TLC vertical launched) anti-submarine ballistic missile element of the Club-N complex included an MPT-1UME high speed homing torpedo that separated from the missile body following the flight to the target area. The torpedo then immersed in the water, thereafter, the onboard sonar target seeker would guide the weapon to the target.

For engagement of fixed location land targets the Club-N complex could also be armed with the 3M-14E/TE LACM (Land Attack Cruise Missile), which incorporated a booster section, a low-altitude supersonic sustainer section and, in the 3M-14TE, an integral TLC for vertical launch (Rosoboronexport).

Mock-up of a notional launch complex with representative shapes of the various weapons of the Club-N missile complex. Rosoboronexport

The Club-N complex automated fire control system operated in real-time, utilising target data provided by the ship sensors through the combat information management system, or by direct manual input and through the input of navigational data from the shipborne navigation suite. The anti-ship missile variants were equipped with a modern INavS (Inertial Navigation System) to guide them to the target area where a jam-resistant active radar seeker took over for the terminal phase of the flight. The 91RE1/2 anti-submarine ballistic missile element of the Club-N complex was guided to the target area by the onboard INavS, the MPT-1UME high-speed homing torpedo, after separation and immersion in the water, being guided to the target by an onboard sonar. The various components of the system were completely waterproof/flash fire proof, with no requirements for cooling (Rosoboronexport).

Technical Specification Club-N Anti-Ship/Anti-Submarine Missiles

	3M-54TE	3M-54TE1	91RET1
Length:	8.916 m	8.916 m	8.916 m
Diameter:	0.645 m	0.645 m	0.645 m
Weight:	3655 kg	3210 kg	3105 kg
Warhead weight:	200 kg	400 kg	500 kg
Range:	Up to 220 km		Up to 40 km
Sustainer phase speed:	Mach 0.6-0.8 (up to Mach 3 180-240 m/s in terminal phase)	Mach 0.6-0.8	Mach 2.0 (830 m/s)
Guidance system:	Inertial + active radar seeker		United inertial + acoustic target seeker

*   91RET1 characteristics representative of the 91RE2

3M-14TE Land Attack Cruise Missile Specification – Rosoboronexport

Length: 8.916 m
Diameter: 0.645 m
Weight: 3150 kg
Warhead weight: 450 kg
Guidance: Doppler/inertial + inertial GPS/GLONASS (Global Positioning System/Globanaya Navigozionnaya Sputnikovaya Sistema) navigation receiver

Graphic depicting a quartet of Club-N missiles in flight. Rosoboronexport

The Russian domestic Project 11356M fleet was armed with the Kalibr-NK, which had inherent capabilities significantly in advance of the Club-N. The Kalibr-NK complex was armed with the 3M-14T LACM developed by JSC Experimental Machine Design Bureau Novator. This missile had an effective range of ~2000 km, bestowing upon the operator a powerful stand-off strike capability against fixed position land targets. Kalibr would cruise to the target, under the power of a single NPO Saturn 36MT turbofan engine, in complex flight profiles, determined by a number of factors. These included the nature of the defences and terrain to be overflown, at altitudes down to about 30 m, the missile having an accuracy of around 5 m. The target would be destroyed by a 500 kg class high explosive warhead (Harkins, 2016S). Kalibr could be armed with a conventional or nuclear warhead.

«KALIBR»
Sea-based cruise missiles

Launch platform:	naval ships, submarines
Firing range:	approximately 2,000 km
Operational altitude:	30 m and more
Flight profile:	changable (depends on terrain)
Warhead weight:	approximately 500 kg
Warhead type:	conventional and nuclear
Probable miss distance:	approximately 5 m

The Kalibr missiles arming Russian Navy Project 11356M had significantly enhanced capabilities over those of the Club-N complex, most notable of which was the extended firing range of around 2000 km and more powerful warhead – nuclear or conventional in the land attack and possibly anti-ship variants. MODRF

OKB Novator Kalibr-NK 3M-14T/3M-14

Engine: One NPO Saturn 36MT turbofan rated at 450 kgf maximum thrust
Range: ~2000 km
Operational altitude: 30 m and upwards
Flight profile: Dependent upon terrain to be overflown
Warhead type: Conventional and nuclear
Warhead weight: 500 kg class
Accuracy: Within 5 m
Launch platforms: Surface warships (3M-14T) and submarines (3M-14)

Top: Graphic depicting a Kalibr missile homing on a target just above the wave-tops of a rough sea. Bottom: The Project 11356M Guard Ship, *Admiral Essen* of the Black Sea Fleet, launches a Kalibr-NK 3M-14T land attack cruise missile from the vertical launch complex in the forward section of the ship, just aft of the Shtil-1 vertical launch complex. MODRF

The second batch of Project 11356 Frigates built for India, the Teg Class (INS *Tarkash* **illustrated), was armed with the Brahmos supersonic anti-ship/land attack cruise missiles complex in place of Club-N.** KB Ametist

BRAHMOS – The second series of Project 11356 Frigates built for the Indian Navy, the Teg Class, was armed with the Brahmos two-stage fire and forget high supersonic speed cruise missile system for ship launch (developed in anti-ship and land attack variants), which could engage targets with high precision at distances out to 290 km from the launch platform. The ship launched missile, like the land launched missile, was housed in an integral TLC (Severnoye, 2012a & DRDO). The first stage consisted of a solid propellant booster that propelled the missile to supersonic speed before separating from the vehicle, propulsion then being taken over by the second stage liquid propellant ramjet that would accelerate the missile to a speed approaching Mach 3 for the cruise phase of the flight. The multi-Mach speed, which was maintained throughout the flight to the target, would reduce target defensive reaction time through a reduction in flight time compared to subsonic missiles and would reduce the potential for the missile to be intercepted by defensive missile/gun systems.

The Brahmos missile could adopt a number of flight profiles at various flight altitudes, up to a maximum of 15000 m, during the cruise to the target. Flight altitude reduced to around 10 m in the terminal phase, the target being destroyed or disabled by the 200-300 kg high explosive warhead. The high supersonic impact speed bestowed upon Brahmos a high kinetic energy impact, calculated at around nine times that of a typical subsonic cruise missile (Brahmos Aerospace).

Brahmos was introduced to Indian naval service (trials) aboard the INS *Rajput* in 2005. It was adopted as the standard anti-ship cruise missile for new acquisition Indian navy warships, with the intention to retrofit certain existing vessels as they entered modernisation/refit – the second series of Project 11356 built for India, the Teg Class, were armed with eight Brahmos missiles (Brahmos Aerospace) launched from the vertical launch complex formerly occupied by Club-N.

The Brahmos International supersonic cruise missile emerged as an Indian/Russian development of the latter nations Yakhont ship/submarine launched supersonic anti-ship/land attack cruise missile first introduced to service in 1991 when the Russian Federation was a constituent nation of the Soviet Union (dissolved on 25 December 1991). Brahmos was integrated onto the second series of Project 11356 Frigates for the Indian Navy, the Teg Class, and is expected to constitute the main anti-ship/land attack missile armament of the four Project 11356 of the third series under construction/order for the Indian Navy in 2021. Brahmos International/Strela PA

A Brahmos missile is launched from the INS *Ranviyay* during trials. DRDO

The Almaz Kashtan-M ('Chestnut') missile gun complex appears to be something of an enigma in regard to the Project 11356/M. The weapon system is highlighted in Severnoye and Rosoboronexport documentation as being an integral part of the armament suite of the Project 11356 (Severnoye & Rosoboronexport). Furthermore, in a press release of 11 March 2016, the Russian Western Military District described the first Project 11356M for the Russian Navy, *Admiral Grigorovich*, as being the first Project 11356 ship armed with a Kashtan-M anti-aircraft missile/gun system (MODRF Press Service of the Western Military District, 2016f). That said, the system appears not to have been installed on any Project 11356/M ship at build, its association with the warship class being an option rather than a delivered installation. The proposed Kashtan complex installation on the Project 11356/M would incorporate two 3R87-E combat modules and a single 3R-86-1E command (target detection and control) module (Severnoye).

The Kashtan-M, a development of the land based Tunguska short-range (point) air defence system, was intended for engagement of anti-ship missiles, aircraft launched bombs and high and low speed aircraft/helicopters at short range (out to 8-10 km). This system, which also had a secondary role of striking small surface targets, comprised a command module to facilitate target detection and target distribution for engagement by the combat modules. In order to achieve a high degree of accuracy in targeting against an incoming projectile MNIIRE Altair developed a fire control system that incorporated optoelectronic and millimeter-wave radar channels, the latter allowing narrower antenna patterns to be obtained over that of centimeter range radar (Almaz). The combat modules would receive targeting data automatically, facilitating automatic target tracking, engagement and reloading (Nudelman Precision Engineering Design Bureau) – the Project 11356/M could be armed with 64 x 9M311-1E surface to air missiles (32 per combat module) and 6,000 rounds of 30 mm ammunition for the two combat modules (Severnoye & Rosoboronexport).

Graphic depicting the Kashtan-M 3R87-E missile/gun air defence complex (top) and a Kashtan-M module during automatic gun firing (bottom). Rosoboronexport

Kashtan-M (Chestnut) specification – data furnished by Nudelman Precision Engineering Design Bureau with input from KBP Tula

Combat Unit weight without ammunition and power units: 7700 kg
Control system: Optoelectronic/radar
Missile load per combat module: 32 x 9M311-K missiles stored in launch containers in reloading module
Maximum velocity of engaged targets: 1000 m/second
Effective range for the missile/gun armament: 1500-8000 m/4000 m (KBP Tula documentation states 10000 m for the missile armament)
Effective engagement altitude for the missile/gun armament: 5-3500 m/5-3000 m (KBP Tula documentation states 6000 m maximum for the missile armament)
Per module 30 mm round capacity/rate of fire: 1000/10,000 rounds per minute
Muzzle velocity: 890 m/second
Reaction time: 6-8 seconds
Number of simultaneously engaged targets (dependent on the number of combat units): 1-6

Increasingly the Russian Helicopters (Kamov) Ka-27PL anti-submarine helicopter is being referred to as the Ka-28 in Russian Federation service. This prefix, previously allocated only to export machines, is increasingly appearing on stenciling on the forward fuselage sections of newer/modernised machines in Russian Naval Aviation service, although the Ka-27PL designation is still utilised. Ilyushin

The Project 11356/M ships were built with an integral hanger and flight deck located in the ship aft section. This facilitated the operation of a single medium size helicopter – Ka-27PL/28 ASW (Anti-Submarine Warfare) or Ka-31 AWACS (Airborne Warning and Control System) helicopter, with support equipment, fueling and, in the case of the Ka-27/28, armament installation infrastructure. The extended Kamov (incorporated within Russian Helicopters) family of Ka-27/28/29/31/32 helicopters – the basic utility configured Ka-27 prototype had conducted its maiden flight in the early 1970's and the initial service variant was introduced in 1981 – covers a number of variants optimised for ASW, search and rescue, assault transport/attack, radar picket (AWACS) and general purpose duties. The main production centre was the Kumertau Aviation Production Enterprise (KumAPE), remaining so in the second decade of the twenty first century. The Ka-27 family would, like its Ka-25 predecessor, feature a co-axial rotor system, the benefits of which, in addition to smaller dimensions due to the removal of the requirement for a tail rotor, included the facilitation of less-demanding deck landings, particularly when encountering crosswinds.

In 2021, the Ka-27PL was the standard Russian ASW helicopter operating from warships of the various Russian Fleets and was operated by the Indian Navy under the Ka-28 or Ka-27 designations. The Ka-27PL was tasked with the detection of modern surface and low acoustic signature sub-surface targets, the data then being relayed to other assets for attack or, alternatively, the target could be directly attacked by the helicopter employing ASW weapons, primarily homing torpedoes.

Ka-27PL of the Naval Aviation of the Russian Black Sea Fleet based in Crimea.
MODRF

The Ka-27 can conduct day/night anti-submarine search missions up to 200 km from the host ship platform in adverse weather conditions and against waves up to 5 points. The Ka-27 can be armed with PLAB 250-120 bombs, various unguided rocket combinations and torpedoes, AT-1M VTT-1, UMGT-1 Orlan, APR-2 Yastreb-M and AT-1M. In its primary ASW mission, the weapon suite allows the Ka-27PL to attack and destroy sub-surface targets moving at speeds up to 75 km/h at depths down to 500 m (MODRF).

A pair of Ka-27PL anti-submarine warfare helicopters of the Russian Black Sea Fleet, with dipping sonar extended during a training operation on 23 May 2020. MODRF

The rubric of the Ka-31, which was developed from the Ka-29 assault transport helicopter, was to provide radar surveillance cover for naval operations beyond the coverage of land based radar or land based airborne radar coverage from AWACS type aircraft – in 2020, the principal AWACS type aircraft in Russian Federation Aerospace Forces and Indian Air Force service was the Beriev A-50. The Ka-31's installed RTK radar complex, which incorporated a powerful radar system and a large rotating antenna, bestowing 360° coverage, would be capable of detecting a multitude of target types, including cruise missiles flying at various altitudes over large distances (Rostec Corporation). This would incorporate the ability to detect very low-altitude targets over the sea surface or land mass against sea and land/surface background clutter. Detected targets would be automatically categorised and identified. The parameters and coordinates of the target trajectory would be identified and the data would then be automatically processed and passed to various recipients. The number of simultaneously tracked targets remained classified in 2021, but was described as 'a large number' (Rostec Corporation). Surface targets could also be detected, the data for the various airborne and surface targets being relayed to ship, ground or airborne control stations and air defence systems, shipborne, airborne or land-based.

The Ka-31, the prototype of which had conducted its maiden flight in 1987, served with the Soviet and later Russian Federation Navies (the latter continued to operate Ka-31R in 2021). India received a total of fourteen Ka-31 helicopters, which could be operated from various Indian Navy warships, but predominantly formed elements of aircraft carrier air groups.

The Ka-27 was powered by two Klimov TV3-117VK (Kamov Marine) engines. Some Ka-27 variants, the Ka-29 and civil Ka-32 helicopters, were powered by the TV3-117VK (high altitude Kamov) engines. Ka-28 export variants of the Ka-27 were powered by the TV3-117VKR (high altitude Kamov Power) engine. Ka-27 and Ka-31 helicopters were later powered by TV3-117VMA (high altitude Modernised Model 'A') engines, which had been developed for the Kamov Ka-50 'Black Shark' attack helicopter, but also powered non Kamov designs, such as Mi-24 models and the Mi-28A/N attack helicopter, serial production commencing in 1986.

A Ka-31R of the Naval Aviation of the Russian Black Sea Fleet during take-off/landing trials aboard a surface ship on 29 February 2020. The Project 11356M can accommodate a Ka-31R airborne warning and control helicopter to provide early warning of an emerging threat, incoming attack or to enhance the overall tactical picture of the air, sea and or land environment. MODRF

Ka-27 Specification – data furnished by Russian Helicopters with input from the MODRF

Powerplant: 2 x TV3-117VK (Russian service), each rated at around 2200 hp.
Length: 12.3 m
Height: 5.4 m
Rotor span: 15.9 m
Empty weight: 6100 kg
Maximum take-off weight: 12000 kg (Russian Helicopters) or 12500 kg (MODRF)
Maximum weight of underslung load: 3775 kg
Maximum flight speed: 250 km/h (Russian Helicopters) or 270 km/h (MODRF)
Maximum ceiling: 5000 m
Flight range: 900 km
Rate of climb: 9.5 m/second
Armament: PLAB 250-120 bomb, AT-1M, VTT-1, UMGT-1, ME Orlan, APR-2 Yastreb-M or ATM-1 torpedoes
Crew: Three (search and attack mission)

Ka-31 Specification – data furnished by Russian Helicopters

Powerplant: 2 x TV3-117VMA, each rated at 2200 hp.
Maximum take-off weight: 12500 kg
Maximum speed: 220 km/h
Operating height of flight within radar surveillance coverage: 1500-3500 m
Maximum ceiling: 5000 m
Flight range: 680 km
Rate of climb: 9.5-12 m/second
Radar equipment: RTK complex
Crew: 2

Rear on port quarter view of the Project 11356M Guard Ship _Admiral Essen_, showing to advantage the hanger complex forward of the flight deck. MODRF

Project 11356M Guard Ships *Admiral Makarov* **(top) and** *Admiral Essen* **(bottom) of the Russian Black Sea Fleet.** MODRF

3

INDIAN SENTINEL/BLACK SEA GUARDIAN

As briefly noted in chapter 1, the first three Project 11356 General Purpose Frigates built for India were delivered in 2003-2004 and the second batch of three was delivered in 2012-2013. The first batch of three Project 11356 built for India was designated Talwar Class. INS (Indian Naval Ship) *Talwar* (F40) was commissioned on 18 June 2003, INS *Trishul* (F43) was commissioned on 25 June 2003 and INS *Tabar* (F44) was commissioned on 19 April 2004. The second batch of three Project 11356 built for the Indian Navy was designated Teg Class, which was equipped with an updated multi-function combat suite, at the heart of which was the incorporation of the Brahmos supersonic surface to surface cruise missile complex as an alternative to the Club-N missile complex arming the Talwar Class (Brahmos Aerospace, 2012). The first of these ships, INS *Teg* (F45), was laid down in July 2007 (Brahmos Aerospace, 2012) and commissioned on 27 April 2012. INS *Tarkash* (F50) was launched at the Yantar Shipyard, Kaliningrad, on 24 June 2010 (Severnoye, 2010) and was commissioned at the Yantar shipyard on 9 November 2012 (Severnoye, 2012b). Indian Navy documentation suggests INS *Tarkash* was commissioned on 12 November 2012 and INS *Trikand* (F51), the third and last of the three second series Project 11356, was formally accepted by the Indian Navy at the Yantar shipyard on 29 June 2013, on which date the Russian flag, under which the vessel had conducted her trials, was lowered and the Indian flag raised. The Teg Class would be based at Mumbai (Indian Navy, 2020, Severnoye, 2013 & Severnoye, 2012a).

In 2006, Severnoye Design Bureau partnered with JSC Yuzhno-Sakhalinsk to develop the Project 11356 to meet a Russian Guard Ship (General Purpose Frigate) requirement. Six Project 11356M were ordered for the Russian Navy under two state contracts from 2009 (MODRF Press Service of the Western Military District, 2016a). All six Project 11356M ships were expected to be delivered to the Black Sea Fleet (Severnoye, 2014). The first of these ships, *Admiral Grigorovich*, was laid down at the Yantar Shipyard, Kaliningrad, on 18 December 2010, the second, *Admiral Essen*, was laid down at the Yantar Shipyard on 8 July 2011 and the third, *Admiral Makarov*, was laid down at the Yantar Shipyard on 29 February 2012 (Severnoye, 2012).

Top: The Project 11356 Frigate for the Indian Navy incorporated many subtle differences from that of the later Project 11356M ships that would be procured for the Russian Black Sea Fleet. The most outwardly apparent difference was the Indian ships single-rail launch system for the Shtil-1 air defence missile complex – the domestic Russian ships having the 3S90 underdeck vertical launch system. Rosoboronexport/USC

Top: INS *Talwar* (F40), the first Project 11356 ship of what would be designated Talwar Class, was commissioned on 18 June 2003. The ship conducted trials under the Russian Flag, with trials side codes. Above: The Talwar Class (INS *Talwar* (F40) illustrated) significantly enhanced the Indian Naval power in its area of responsibility through the enhanced capabilities of the design, in particular in regard to the anti-air warfare and long range anti-ship/land attack missions. SDB/Indian Gov.

The first of the second series of the Project 11356 built for the Indian Navy, INS *Teg*, was laid down in July 2007 and commissioned on 27 April 2012. SDB

The first series of Project 11356 Frigates built for the Indian Navy (INS *Trishful* (F43) top) and the second Project 11356 series (INS *Tarkash* (F50) bottom) were outwardly almost identical. Rosoboronexport/Indian Gov.

The penultimate ship of the second Project 11356 series (Teg Class) INS *Tarkash* launched at the Yantar Shipyard, Kaliningrad, on 24 June 2010. SDB

The last Project 11356 Frigate of the second series built for India, INS *Trikand* (F51), was commissioned at the Yantar Shipyard, Kaliningrad, on 29 June 2013. SDB

The lead Project 11356M Guard Ship (classified as a pre-series ship) for the Russian Navy, *Admiral Grigorovich*, was launched at the Yantar Shipyard, Kaliningrad, on 14 March 2014 (MODRF Press Service of the Western Military District, 2016a). The ship entered state acceptance trials around 28 October 2015 (Severnoye, 2015) and arrived at the Northern Fleet base, Severomorsk, for weapon trials under state sea trials on 21 December that year (MODRF Press Service of the Northern Fleet, 2015). Following conclusion of state acceptance/sea trials, the *Admiral Grigorovich* was commissioned in a flag raising ceremony at the Yantar shipyard on 11 March 2016 (MODRF Press Service of the Western Military District, 2016a & MODRF Press Service of the Western Military District, 2016f).

The *Admiral Essen* (first serial Project 11356M for the Russian Navy) was launched on 7 November 2014 and commenced factory sea trials in the Baltic Sea on 28 October 2015 (MODRF Press Service of the Western Military District, 2016a). The *Admiral Essen* arrived at Kronstadt, Russia, on 21 March 2016, on a transit that afforded the opportunity to test the ships high-speed and maneuvering capabilities and ship systems. Following replenishment, the ship was transferred to the Northern Fleet for another phase of state trials/armament trials, arriving at, Severomorsk, Russia, for armament firing trials on 4 April 2016, after which the ship would return to Kaliningrad to commission into Russian naval service (MODRF Press Service of the Western Military District, 2016a & Information Activities Office of the Northern Region, Severomorsk, 2016). Following completion of state acceptance tests the ship sailed for Baltiysk to commence the process of final acceptance by the MODRF (Ministry of Defence of the Russian Federation) (Severnoye, 2016).

Page 68-69: The lead Project 11356M Frigate (Guard Ship) for the Russian Black Sea Fleet, *Admiral Grigorovich*, **was launched at the Yantar Shipyard, Kaliningrad, on 14 March 2014. The Project 11356M for Russian domestic service took on an overall similar outward appearance to the Talwar Class and Teg Class built for India, but had notable differences, such as the lack of rail launch system for the Shtil-1 air defence missile complex.** SDB

The trails phases in the Baltic Sea and Northern Fleet areas respectively involved not only sea trials, but also weapon trials and helicopter trials, the latter involving in excess of 50 helicopter landings on the ships flight deck. During the trials the ship travelled in excess of ~32186 km (20,000 miles) (MODRF Press Service of the Western Military District, 2016). Following completion of state acceptance tests the

Admiral Essen was commissioned into Russian Navy service in a flag raising ceremony at the Yantar Shipyard on 7 June 2016. The ship was deployed to Sevastopol, Crimea, to join the Russian Black Sea Fleet for permanent basing on 28 April 2017 (MODRF Press Service of the Southern Military District, 2017f & Severnoye, 2016a).

Стapoжевой корабль проекта 11356
«Адмирал Эссен»
заложен 8 июля 2011 года
в г. Калининграде
на Прибалтийском судостроительном заводе «Янтарь»
спроектирован в г. Санкт-Петербурге Северным ПКБ

Previous page top: The first Project 11356M, *Admiral Grigorovich*, during artillery weapon system trials at sea on 13 July 2016. Previous page bottom: The *Admiral Grigorovich* during state acceptance tests. This page: Plaque commentating the keel laying for the second Project 11356M, *Admiral Essen*, for the Russian Black Sea Fleet (first series build) on 8 July 2011 (top) and the *Admiral Essen* launch on 7 November 2014 (bottom). SDB

Top: The *Admiral Essen* at Kronstadt where it arrived on 16 March 2016, during state acceptance tests. Bottom: The *Admiral Essen* during armament trials with the Russian Northern Fleet in ice covered waters, operating out Severomorsk, where it arrived on 4 April 2016. MODRF

Top: The *Admiral Essen* returned to the Yantar Shipyard, Kaliningrad, following completion of state tests with the Russian Northern Fleet. Bottom: The *Admiral Essen* during trials in the Baltic Sea, circa April/May 2016. SDB/MODRF

Top: The *Admiral Essen* was commissioned into the Russian Navy at the Yantar Shipyard, Kaliningrad, Russia, on 7 June 2016. **Bottom:** The *Admiral Essen* arrived at its permanent base, Sevastopol, Crimea, following transit from the Baltic to the Black Sea in summer 2016. SDB

The third Project 11356M built for the Russian Navy was the SKR *Admiral Makarov*, which was built to a more advanced standard in regard to certain ship systems – radio-technical, weapon and life support – than was her Project 11356M predecessors (Department of Information and Mass Communications of the

Ministry of Defence of the Russian Federation, 2015). The *Admiral Makarov* was launched at the Yantar Shipyard, Kaliningrad, on 2 September 2015 (Department of Information and Mass Communications of the Ministry of Defence of the Russian Federation, 2017a & Severnoye, 2016). At that time, the *Admiral Grigorovich* was undergoing sea trials and the *Admiral Essen* was undergoing factory harbor trials (Severnoye, 2015).

At the time of her launch in 2015, the *Admiral Makarov* crew was being formed (Department of Information and Mass Communications of the Ministry of Defence of the Russian Federation, 2015), the process being completed by 12 January 2016, before crew training commenced in St Petersburg (MODRF Press Service of the Southern Military District, 2016c) in preparation for commencing factory sea trials, which were completed in late (29) November 2016, following armament firing trials in the Barents Sea (Severnoye, 2016a). The *Admiral Makarov* recommenced state testing at the Baltic Fleet Sea ranges on 1 April 2017 (MODRF Press Service of the Western Military District, 2017a) and completed state acceptance tests on 15 September that year. The ship was formally commissioned into the Russian Navy at the Yantar shipyard on 27 December 2017 (Department of Information and Mass Communications of the Ministry of Defence of the Russian Federation, 2017 & Department of Information and Mass Communications of the Ministry of Defence of the Russian Federation, 2017a).

The keel for the third Project 11356M for the Russian Black Sea Fleet, *Admiral Makarov*, was laid at the Yantar Shipyard, Kaliningrad, on 29 February 2012. SDB

Top: Plaque commemorating the keel laying of the *Admiral Makarov* on 29 February 2012. Bottom: The *Admiral Makarov* during sea trails, which were successfully completed on 27 September 2017. SDB

Top: The *Admiral Makarov* on completion of state acceptance tests in November 2017. The *Admiral Makarov* was commissioned into service at the Yantar Shipyard, Kaliningrad, and transferred to the Russian Navy on 25 December 2017. SDB

Keel laying (top) and plaque commemorating the keel laying (bottom) of the fifth Project 11356M originally intended for the Russian Navy, *Admiral Istomin* (Russian Admiral killed in defence of Sevastopol during the Crimean War in 1855), at the Yantar shipyard on 15 November 2013. SDB

The fourth Project 11356M ship intended for the Russian Navy, the *Admiral Butakov*, was laid down on 12 July 2013. The fifth Project 11356M intended for the Russian Navy, *Admiral Istomin*, was laid down at the Yantar shipyard on 15 November 2013. The sixth Project 11356M for the Russian Navy was named *Admiral Komilov*. The six ships had originally been scheduled for delivery to the Russian Navy in the period 2014-2016 (Severnoye, 2013a).

The *Admiral Grigorovich* departed Sevastopol bound for the Mediterranean on 27 February 2017 (top) and *Admiral Essen* returned to Sevastopol on 21 September 2017, following a Mediterranean deployment. MODRF

All three of the Black Sea Fleet Project 11356M ships were heavily involved in rotational deployments under the permanent connection of the Russian Navy in the Mediterranean, including participating in Russian operations in support of Syrian Government forces fighting against extremist/opposition forces occupying parts of that country. This also brought the Russian mission into what could be termed soft operations against NATO (North Atlantic Treaty Organisation) naval forces in the Mediterranean Sea – tracking NATO submarines that were deemed an impediment to the Russian operation against ISIL (Islamic State of Iraq and the Levant) etc.

Under one such Mediterranean deployment, the *Admiral Essen* entered the Mediterranean Sea on 5 May 2017 (MODRF, 2017a). Cruising off the Syrian Coast in the Eastern Mediterranean Sea four days later, the *Admiral Essen*, along with the Black Sea Fleet Project 636.3 submarine *Krasnodar* (submerged), conducted a Kalibr land attack cruise missile strike against ISIL targets near to the city of Deir Ez-Zor in Eastern Syria (MODRF). Following her Mediterranean deployment, the *Admiral Essen* returned to Sevastopol, Crimea, on 5 July 2017 (MODRF, 2017b).

The deployments to the Mediterranean continued through 2020 and were set to be an ongoing element of the permanent connection of the Russian Navy in the Mediterranean, operating out of Tartus, Syria. The Project 11356M ships were also tasked with intermittent Atlantic Ocean deployments, the *Admiral Grigorovich* completing one such cruise when it reentered the Mediterranean Sea through the Strait of Gibraltar on 19 August 2019 (MODRF Press Service of the Southern Military District, 2019e).

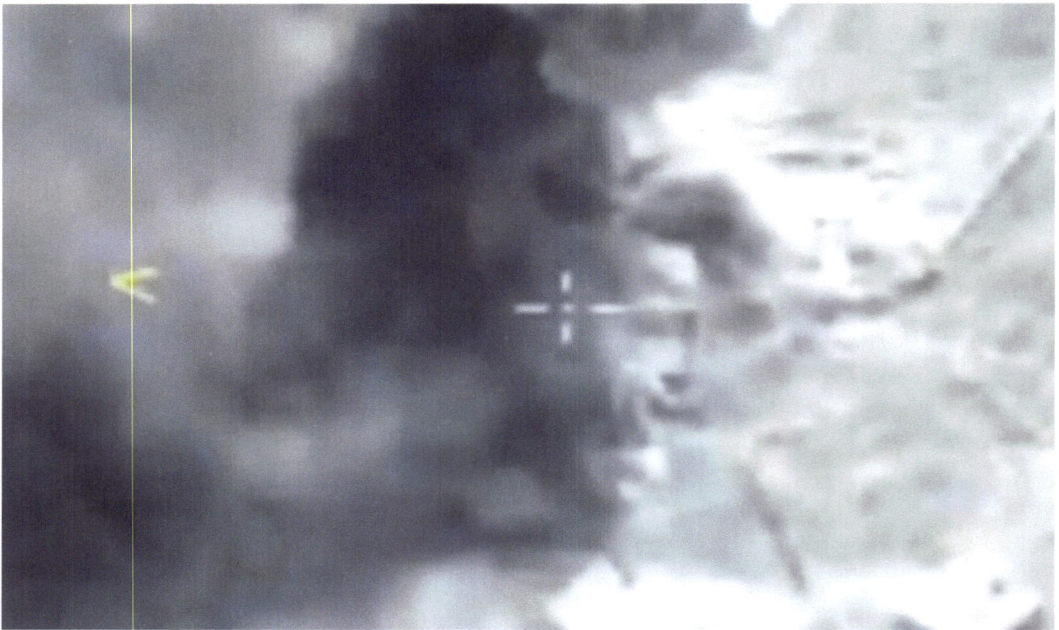

Page 80-81: The *Admiral Essen*, cruising off the Syrian coast on 9 May 2017, launched a salvo of Kalibr-NK land attack cruise missiles against ISIL targets near Deir ez-Zor, Eastern Syria, in support of Syrian Government forces in the fight against the extremist organisation during the Syrian Civil War. This page: A Kalibr-NK cruise missile approaches (from the right of image) the target in Deir ez-Zor (top) and the target in the aftermath of the strike (bottom). MODRF

Project 11356M Guard Ship (Frigate) *Admiral Grigorovich* **of the Black Sea Fleet.**
MODRF/SDB

Project 11356M Guard Ship (Frigate) *Admiral Essen* **of the Russian Black Sea Fleet.**
MODRF

Project 11356M Guard Ship (Frigate) *Admiral Makarov* **of the Black Sea Fleet.**
MODRF

The Project 11356M Guard Ship *Admiral Makarov* participated in the main Russian Naval parade in the Baltic on 29 July 2018. The ship commenced transfer from the Baltic Sea to the Black Sea on 18 August 2018, and arrived in Sevastopol on 5 October that year. MODRF

This page: The Project 11356M ships built for the Russian Black Sea Fleet may, at a future date, be armed with the Tsirkon hypersonic surface to surface missile trialed on the Project 22350 Large General Purpose Frigate *Admiral of the Fleet of the Soviet Union, Gorshkov* **(top). Tsirkon is fired from the standard 3S-14 vertical launch system on the Project 22350 and would be launched from the existing vertical launch complex on the Project 11356M.** MODRF/SDB

The operational capability of the three Project 11356M built for the Russian Black Sea Fleet may be enhanced through adoption of new systems as they become available, including the potential for arming the ships with the Tsirkon (Zircon) hypersonic (around Mach 8-9) anti-ship/land attack cruise missile, which would be launched from the existing vertical launch complex for the Kalibr-NK. In late 2020, this weapon system was undergoing an extensive phase of test firings from the 3S-14 vertical launch complex on the Project 22350 Large General Purpose Frigate, *Admiral of the Fleet of the Soviet Union, Gorshkov.*

In 2021, two Project 11356 ships were under construction at the Yantar Shipyard, Kaliningrad and two more were under order to be built at Goa Shipyard Limited, India, for the Indian Navy. Rosoboronexport

The three ships of the second series ordered by the MODRF for the Russian Navy were not completed due to problems in sourcing suitable power plants. The original power plants had not been delivered by Ukraine following a referendum in Crimea and the Special region of Sevastopol (these regions were previously administered by Ukraine), to secede from Ukraine due to persecution of the Russian speaking population in what has been termed a de-Russification program (removal of to use the Russian language from Russian speaking populations etc.) forced on the population following a 2014 Ukrainian coup that led to a change in government and political direction. Crimea, which, following the referendum and succession, re-unified with Russia, of which it was historically an integral part until gifted to the Ukrainian Soviet Republic (without democratic process) by the Soviet Union in the 1950's, would ultimately become the home base of the Project 11356M Guard Ships operating with the Russian Black Sea Fleet.

India ordered a third series of four Project 11356 in November 2018. Two of these ships would be built at Yantar Shipyard for completion by 2023 and the other two were to be built at the Goa Shipyard Limited in India (Yantar Shipyard, 2020). The two ships to be built at the Yantar shipyard were the first two second series Project 11356M ships, *Admiral Butakov* and *Admiral Istomin* laid down on 12 July 2013 and 15 November 2013 respectively for the Russian Navy. Construction of the last of the six Project 11356M ordered for the Russian Navy, *Admiral Kornilov*, had continued at Yantar Shipyard. In February 2021, the yard submitted a proposal for completion of this ship for the Russian Navy or, alternatively, a foreign customer as the power plant situation for the domestic Russian ships was solved in 2020 through development of Russian gas turbines for ships under the Russian foreign purchase replacement program implemented to counter western sanctions on the country. Russia is unlikely to order additional Project 11356M ships as, in line with evolving operational requirements and reduced defence spending, it pursues other programs of warship building, such as the Project 22350 Large General Purpose Frigate, designed to enhance on the capabilities extant with the Project 11356M.

GLOSSARY

3-D	Three dimensional
arc min	1 arc minute = 0.25°
ASCM	Anti-Ship Cruise Missile
AWACS	Airborne Warning and Control System
C	Centigrade
C2	Command & Control
CIWS	Close In Weapon System
ECM	Electronic Countermeasures
FSUE	Federal State Unitary Enterprise
GLONASS	Globanaya Navigozionnaya Sputnikovaya Sistema (Global Navigation Satellite System)
GPS	Global Positioning System
hp.	Horse power
HUMSA	Hull Mounted Sonar Advanced
INavS	Inertial Navigation System
INS	Indian Naval Ship
ISIL	Islamic State of Iraq and the Levant
kg	Kilogram
kgf	Kilogram force
kHz	Kilohertz
km	Kilometre
km/h	Kilometres per Hour
knot(s)	Nautical Miles per Hour
KumAPE	Kumertau Aviation Production Enterprise
kW	Kilowatt
m	Metre
m^2	Metre squared
m/s	Metres per second
mm	Millimetre
MODRF	Ministry of Defence of the Russian Federation
NATO	North Atlantic Treaty Organisation
RCS	Radar Cross Section
SDB	Severnoye Design Bureau
SMP	State Moscow Plant
USC	United Shipbuilding Corporation
φ	Typically symbol for phi (21st letter of the Greek alphabet)
$\leq$	Less than or equal to
$<$	Strict inequality - Less than
$>$	Strict inequality - Greater than
$\circ$	Degree(s)
$\pm$	Plus or minus
$\sim$	Approximately equal to (can also be used to mean asymptotically equal)

BIBLIOGRAPHY

A large number of documents were consulted in the preparation of this volume. Whilst some could be standard referenced, many diagrams and small-scale pieces of information from design bureau, builders or operators cannot be easily referenced. The major contributors of textual and graphic material include Almaz-Antey, Brahmos Aerospace, Concern Agat, Concern CSRI Elektropribor State Research Centre of the Russian Federation, Concern Okeanpribor, Council of the European Union, CRI Burevestnik, Department of Information and Mass Communications of the Ministry of Defence of the Russian Federation, Defence Research and Development Organisation (DRDO), Federal State Unitary Enterprise State Moscow Plant, Salyut, PJSC Ilyushin Aviation Complex, Information Activities Office of the Northern Region, Severomorsk, JSC Scientific Production Association Alloy, JSC Tulamashzavod Production Association, KB-Ametist, KB Arsenal, Indian Navy (Government of India), KBP Tula, Ministry of Defence of the Russian Federation, MODRF Press Service of the Western Military District, Morinformsystem-Agat Concern, NPO Meridian, Nudelman Precision Engineering Design Bureau, Press Service of the Northern Fleet, Press Service of the Western Military District, Press Service of the Southern Military District, Rosoboronexport, Russian Defence Export, Russian Helicopters, Severnoye (Severnoe) Design Bureau (SDB), Strela PA, United Ship Building Corporation (USC) and Yantar Shipyard.

Council of the European Union (2009) 'Independent International Fact Finding Mission on the Conflict in Georgia', Volume 1, Council of the European Union, Brussels

Harkins, H (2016) 'Russian Non-Nuclear Attack Submarines' *Project 877/877E/877EKM/Project 636/636.3 & Project 677/Amur 1650/950/S-1000*, Centurion Publishing, United Kingdom

Harkins, H (2017) 'Russian/Soviet Aircraft Carrier & Carrier Aviation Design & Evolution Volume 2', *Aircraft Carrying Heavy Cruisers – Project 1143.5/6 Kuznetsov Class, INS Vikramaditya & the Nuclear Powered ASW/Attack Carriers – Project 1153, Project 1160, Project 1143.7 Ul'yanovsk & Project 23000E*, Centurion Publishing, United Kingdom

Almaz-Antey (2019) *Shtil-1 multi-channel single-rail launch medium range surface to air missile system*, Almaz-Antey, Russia

Brahmos Aerospace (2020) *Ship-based weapon complex*, Brahmos Aerospace Private Ltd, New Delhi, India

Concern Agat AK-630M complex specification, Concern Agat, Russia

CRI Burevestnik (2020) *A-190-01 gun complex* detail, CRI Burevestnik, Russia

CSRI Elektropribor (2013) *Inertial Navigation and Stabilization system Ladoga-ME*, Concern CSRI Elektropribor, State Research Centre of the Russian Federation

Ametist (2016) *5P-10-03E, Universal small-sized radar fire control system of naval artillery*, KB-Ametist, Moscow, Russia

FSUE SMP Salyut (undated) *3-D Radar Fregat-M2EM*, Federal State Unitary Enterprise State Moscow Plant, Salyut, Moscow, Russia

KB Arsenal (circa 2016) *RBU-6000* Product sheet, JSC KB Arsenal, Russia

KB Arsenal (circa 2016) *A-190* Product sheet, JSC KB Arsenal, Russia

MODRF (2015) *Patrol ship 'Admiral Grigorovich' arrived at the Northern Fleet to test weapons*, Ministry of Defence of the Russian Federation, Press Service of the Southern Military District, 12.21.2015

MODRF (2015) *The newest patrol ship 'Admiral Makarov' launched*, Department of Information and Mass Communications of the Ministry of Defence of the Russian Federation, 09.02.2015

MODRF (2016) *Admiral Essen guard ship arrived at the Northern Fleet*, Ministry of Defence of the Russian Federation, Information Activities Office of the Northern region (Severomorsk), 04.04 2016

MODRF (2016) *For the first time the St. Andrews flag was raised on the patrol ship 'Admiral Essen'* Ministry of Defence of the Russian Federation, Press Service of the Western Military District, 06.07.2016

MODRF (2016) *New patrol ship 'Admiral Essen' arrived in Kronstadt as part of state tests, Ministry of Defence of the Russian Federation,* Press Service of the Southern Military District, 03.21.2016

MODRF (2016) *St. Andrews flag was raised on the new patrol ship 'Admiral Grigorovich'*, Ministry of Defence of the Russian Federation, Press Service of the Western Military District, 11.03.2016

MODRF (2017) *Admiral Essen newest frigate of the Black Sea Fleet entered the Mediterranean Sea*, Ministry of Defence of the Russian Federation, 5 May 2017

MODRF (2017) *Admiral Grigorovich frigate of the Black Sea Fleet arrived in Novorossiysk*, Ministry of Defence of the Russian Federation

MODRF (2017) *Black Sea Fleet's frigate Admiral Essen arrived in Sevastopol*, Ministry of Defence of the Russian Federation, Press Service of the Southern Military District, 05.07.2017

MODRF (2017) *Black Sea frigate Admiral Essen returns to Sevastopol from Mediterranean*, Ministry of Defence of the Russian Federation, Press Service of the Southern Military District, 21 September 2017

MODRF (2017) *Frigate of the Black Sea Fleet Admiral Grigorovich arrived in Sevastopol after completing tasks in the Mediterranean Sea*, Ministry of Defence of the Russian Federation, Press Service of the Southern Military District, 10.04.2017

MODRF (2017) *Frigate of the Black Sea Fleet Admiral Grigorovich will join the Russian Navy in the Mediterranean Sea*, Ministry of Defence of the Russian Federation, Press Service of the Southern Military District, 12.01.2017

MODRF (2017) *New frigate of the Black Sea Fleet 'Admiral Essen' arrived in Sevastopol*, Ministry of Defence of the Russian Federation, Press Service of the Southern Military District, 07.05.2017

MODRF (2017) *The frigate 'Admiral Makarov' will continue state tests at sea ranges of the Baltic Fleet*, Ministry of Defence of the Russian Federation, Press Service of the Western Military District 05.11.2018

MODRF (2017) *The Navy will include the frigate 'Admiral Makarov' of project 1135.6*, Department of Information and Mass Communications of the Ministry of Defence of the Russian Federation), 09.15.2017

MODRF (2017) *The new Black Sea Frigate 'Admiral Essen' arrived in Sevastopol*, Ministry of Defence of the Russian Federation, Press Service of the Southern Military District, 05.07.2017

MODRF (2017) *The newest frigate 'Admiral Essen' built for the Black Sea Fleet, went to the place of permanent basing*, Ministry of Defence of the Russian Federation, Press Service of the Southern Military District, 02.28.2017

MODRF (2017) *The newest frigate 'Admiral Essen' of the Black Sea Fleet has arrived in the Mediterranean Sea*, Ministry of Defence of the Russian Federation, Press Service of the Southern Military District, 05.05.2017

MODRF (2017) *The newest frigate 'Admiral Makarov' will host a solemn raising of the Andreevsky flag and the ship will be accepted into the Navy*, Department of Information and Mass Communications of the Ministry of Defence of the Russian Federation, 12.26.2017

MODRF (2018) *The frigate of the Black Sea Fleet 'Admiral Makarov' left Sevastopol and headed for the Mediterranean Sea*, Ministry of Defence of the Russian Federation, Press Service of the Southern Military District, 05.11.2018

MODRF (2018) *The newest frigate 'Admiral Makarov' began the transition to the Black Sea Fleet*, Ministry of Defence of the Russian Federation, Press Service of the Southern Military District, 18.08.2018

MODRF (2018) *The newest frigate of the Black Sea Fleet 'Admiral Makarov' first arrived in Sevastopol*, Ministry of Defence of the Russian Federation, Press Service of the Southern Military District, 15.10.2018

MODRF (2019) *The crew of the Frigate Admiral Essen performs firing at sea and air targets in the Black Sea*, Ministry of Defence of the Russian Federation, Press Service of the Southern Military District, 10.12.2019

MODRF (2019) *The frigate of the Black Sea 'Admiral Grigorovich' passed the Strait of Gibraltar and returns to Sevastopol*, Ministry of Defence of the Russian Federation, Press Service of the Southern Military District, 08.18.2019

MODRF (2020) *In the Mediterranean Sea, the Black Sea Fleet frigate 'Admiral Essen' conducted an exercise with a Ka-27PL*, Ministry of Defence of the Russian Federation, Press Service of the Southern Military District, 11.11.2020

MODRF (2020) *The frigate 'Admiral Essen' made a call at the Russian Navy base in the Syrian Tartus*, Ministry of Defence of the Russian Federation, 09.10.2020

MODRF (2020) *The frigate 'Admiral Essen' of the Black Sea Fleet passes the Black Sea Straits in the direction of the Mediterranean Sea*, Ministry of Defence of the Russian Federation, Press Service of the Southern Military District, 29.09.2020

Okeanpribor *MGK-335-03* detail document, OJSC Concern Okeanpribor, St. Petersburg, Russia

Okeanpribor *Vinyetka-EM* detail document, OJSC Concern Okeanpribor, St. Petersburg, Russia

Okeanpribor *Vinyetka-EM-01* detail document, OJSC Concern Okeanpribor, St. Petersburg, Russia

Rosoboronexport (undated) *11356*, Rosoboronexport, Russian Defence Export, Moscow, Russia

Rosoboronexport (undated) *3Ts-25E*, Rosoboronexport, Russian Defence Export, Moscow, Russia

Rosoboronexport (undated) *A-190E-5P-10E*, Rosoboronexport, Russian Defence Export, Moscow, Russia

Rosoboronexport (undated) *AK-630M*, Rosoboronexport, Russian Defence Export, Moscow, Russia

Rosoboronexport (undated) *Club-N*, Rosoboronexport, Russian Defence Export, Moscow, Russia

Rosoboronexport (undated) *Ka-31*, Rosoboronexport, Russian Defence Export, Moscow, Russia

Rosoboronexport (undated) *PK-10*, Rosoboronexport, Russian Defence Export, Moscow, Russia

Rosoboronexport (undated) *Shtil-1*, Rosoboronexport, Russian Defence Export, Moscow, Russia

Severnoye (2010) *Another Frigate for the Indian Navy was launched in Kaliningrad*, Severnoye Design Bureau, 28 June 2010

Severnoye (2010) *New Guard Ship of Project 11356 for the Russian Navy*, Severnoye Design Bureau, 20 December 2010

Severnoye (2012) *Commissioning Ceremony for Project 11356 INS*, Severnoye Design Bureau, 15 November 2012

Severnoye (2012) *Commissioning of INS Teg*, Severnoye Design Bureau, 3 May 2012

Severnoye (2012) *Solemn Keel Laying of the Third Project 11356 Escort Ship*, Severnoye Design Bureau, 2 March 2012

Severnoye (2013) *On November 15 2013 the solemn keel-laying ceremony for the fifth ship in the series of the Project 11356 took place in the territory of the Kaliningrad based Yantar Shipyard*, Severnoye Design Bureau, 20 November 2013

Severnoye (2013) *Solemn delivery of Last Ship of the Second Series of Project 11356*, Severnoye Design Bureau, 3 July 2013

Severnoye (2014) *Launching ceremony of the lead ship of Project 11356 Admiral Grigorovich at the Yantar Shipyard*, Severnoye Design Bureau, 20 March 2014

Severnoye (2015) *Admiral Makarov Frigate Launched in Kaliningrad*, Severnoye Design Bureau, 15 September 2015

Severnoye (2015) *State Acceptance Trials of Admiral Grigorovich*, Severnoye Design Bureau, 28 October 2015

Severnoye (2016) *Admiral Makarov Completed Trials*, Severnoye Design Bureau, 29 November 2016

Severnoye (2016) *Frigate Admiral Essen Completed SAT,* Severnoye Design Bureau, 19 April 2016

Severnoye (2016) *Newest Frigate Admiral Essen Enters Russian Navy Service*, Severnoye Design Bureau, 7 June 2016

Severnoye (2017) *Project 11356 Frigate* product documentation, JSC Severnoye Design Bureau, St Petersburg, Russia

Severnoye (2020) *History*, JSC Severnoye Design Bureau, St Petersburg, Russia

Severnoye (2021) *Project 11356 Frigate* product documentation, JSC Severnoye Design Bureau, St Petersburg, Russia

Splav (2016) *RBU-6000* detail specification sheet, JSC Scientific and Production Association Splav, Russia

Talwar Class Indian Navy Detail Sheet, Indian Government

TASS (2021) *Russian Shipyard submits proposal on completing construction of Project 11356* Frigate, TASS News Agency

Teg Class Indian Navy Detail Sheet, Indian Government

Yantar Shipyard (2020) *History*, Yantar Shipyard, Kaliningrad, Russia

Tulamashzavod AK-630M complex specification, JSC Tulamashzavod Production Association

ABOUT THE AUTHOR

Hugh Harkins FRAS, MIstP, MRAeS is a physicist/historian and author with an extensive research/study background in aeronautic, astronautic, astrophysics, geophysics, nautical and the wider scientific, technical and historical fields. He is also involved in research in the field of Scottish history, which formed a significant element of dual undergraduate degrees. Hugh has published in excess of seventy books, non-fiction and fiction, writing under his given name as well as utilising several pseudonyms. He has also written for several international magazines, whilst his work has been used as reference for many other projects, ranging from the aviation industry, international news corporations and film media to encyclopaedias, museum exhibits and the computer gaming industry. Hugh is an elected Fellow of the Royal Astronomical Society and is an elected member of the Institute of Physics and Royal Aeronautical Society. He currently resides in his native Scotland. Other titles by the author include:

Russian/Soviet Aircraft Carrier & Carrier Aviation Design & Evolution Volume 1 - Seaplane Carriers, Project 71/72, Graf Zeppelin, Project 1123 ASW Cruiser & Project 1143-1143.4 Heavy Aircraft Carrying Cruiser
Soviet Mixed Power Experimental Fighter Aircraft – Piston-Liquid Propellant Rocket Engine/Piston-Ramjet/Piston-Pulsejet & Piston-Compressor Jet Engine Designs of the 1940's
Raid on the Forth - The First German Air Raid on Great Britain in World War II
Light Battle Cruisers and the Second Battle of Heligoland Bight
Russia's Coastal Missile Shield - Bal-E & Bastion Mobile Coastal Cruise Missile Complexes
Iskander - Mobile Tactical Aero-Ballistic/Cruise Missile Complex
Orbital/Fractional Orbit Bombardment System - The Soviet Globalnaya Raketa
Counter-Space Defence Co-Orbital Satellite Fighter
Russia's Strategic Missile Carrier/Bomber Roadmap 2018-2040 – PAK DA, Tu-160M2, Tu-95MSM & Tu-22M3M
Sukhoi T-50/PAK FA - Russia's 5th Generation 'Stealth' Fighter
Sukhoi Su-35S 'Flanker' E - Russia's 4++ Generation Super-Manoeuvrability Fighter
Sukhoi Su-30MKK/MK2/M2 - Russo Kitashiy Striker from Amur
MiG-35/D 'Fulcrum' F – Towards the Fifth Generation
Air War over Syria, Tu-160, Tu-95MS & Tu-22M3 - Cruise Missile and Bombing Strikes on Syria, November 2015-February 2016
Sukhoi Su-27SM(3)/SKM
X-35 – Progenitor to the F-35 Lightning II
X-32 - The Boeing Joint Strike Fighter
Boeing X-36 Tailless Agility Flight Research Aircraft
XF-103 – Mach 3 Stratospheric Interceptor Concept
North American F-108 Rapier - Mach 3 Interceptor
Convair YB-60 - Fort Worth Overcast
Into The Cauldron - The Lancaster MK.I Daylight Raid on Augsburg
Hurricane IIB Combat Log - 151 Wing RAF, North Russia 1941
RAF Meteor Jet Fighters in World War II, an Operational Log
Typhoon IA/B Combat Log - Operation Jubilee, August 1942
Defiant MK.I Combat Log - Fighter Command, May-September 1940
Blenheim MK.IF Combat Log - Fighter Command Day Fighter Sweeps/Night Interceptions, September 1939 - June 1940
Fortress MK.I Combat Log - Bomber Command High Altitude Bombing Operations, July-September1941

www.ingramcontent.com/pod-product-compliance
Lightning Source LLC
Chambersburg PA
CBHW041454210326
41599CB00005B/249